Tree and Forest Measurement

P.W. West

Tree and Forest Measurement

Third Edition

With 34 Figures and 9 Tables

 Springer

P.W. West
School of Environment, Science and Engineering
Southern Cross University
Lismore
New South Wales 2480
Australia

ISBN 978-3-319-14707-9   ISBN 978-3-319-14708-6   (eBook)
DOI 10.1007/978-3-319-14708-6

Library of Congress Control Number: 2015934138

Springer Cham Heidelberg New York Dordrecht London

Cover Photo: A native forest stand of the eucalypt species spotted gum (*Corymbia maculata*) in northern New South Wales, Australia

Printed on acid-free paper

Springer International Publishing AG Switzerland is part of Springer Science+Business Media (www.springer.com)

To Mickie for sharing my life

Preface

Traditionally, forest measurement textbooks have been concerned primarily with determination of the amounts of the principal commercial product of forests, wood. This was certainly the emphasis of the first edition of this book when it was published in 2004. However, since concerns have arisen about global warming, there has been enormous interest in measuring the amount of carbon dioxide that is absorbed from the atmosphere by forests and stored in their biomass. A considerable aid in this has been the rapid development of instruments to measure forests remotely at scales from individual trees on the ground to large-scale images of forests from satellites. Rapid developments in these areas have led to substantial revisions in this edition of the book.

The aim of the book remains to present an introduction to the practice and techniques of tree and forest measurement. It should serve the forestry student adequately in the undergraduate years and be useful as a guide in his or her subsequent professional life. It should allow practising professional foresters to keep themselves abreast of new developments. It aims also to be accessible to landholders and farmers who own and manage forest on their properties, but have no formal forestry education; they may be able to take basic forest measurements themselves and understand the principles of more advanced measurements that professionals take for them.

I have continued to discuss the biological principles that lead to many of the measurements that are made in forests. I believe this will help readers appreciate better why emphasis is placed on measurement of particular things in forests.

Lismore, NSW, Australia   P.W. West
December 2014

Contents

Chapter 1
Introduction

Abstract Measurement of trees and forests is fundamental to the practice of forestry and forest science. Measurements are used to understand how forests grow and develop, to determine how much they contain of the products man wants from them and to ensure that they are managed appropriately. This book is concerned principally with the measurement of the trees in forests, rather than the many other characteristics of forest ecosystems that might be measured. In turn, the book concentrates on the measurement of the amounts of wood tree stems contain and the biomasses of the various parts of trees from their roots up to their leaves. It considers measurements at three scales, from individual trees through groups of trees (stands) up to inventory of large forest areas. It discusses measurement techniques that range from those using simple, hand-held instruments up to those using imagery obtained from aircraft or satellites.

1.1 This Book

The measurement of trees and forests is fundamental to the practice of forestry and forest science throughout the world. Measurements are used to understand how forests grow and develop, to determine how much they contain of the products man wants from them and to ensure that they are managed appropriately.

This book introduces the techniques of tree and forest measurement (or 'mensuration' as it is called in forestry). It covers little more than what might be taught in one semester of an undergraduate forestry course. It should be useful for students and practising foresters as well as for private landholders who own forest and wish either to measure it or understand what professionals are doing when they measure it for them. The book is designed also to assist scientists from other than forestry disciplines who work in forests and need to measure them, although their interests are not necessarily in the trees themselves. It should assist them to take measurements that are consistent with, and comparable to, those that forest scientists have accumulated over many years.

Many of the things that foresters need to know about trees or forests are difficult to measure directly. For example, it is not easy to determine the amount of wood in

© Springer International Publishing Switzerland 2015
P.W. West, *Tree and Forest Measurement*, 3rd edition,
DOI 10.1007/978-3-319-14708-6_1

the stem of a tree standing in a forest, simply because the tree is so tall and large. To deal with this, techniques have been developed to estimate those difficult things from simple measurements that can be taken from the ground. Much of this book is concerned with describing those techniques and how they are applied. However, it does not discuss in any detail how forestry scientists go about developing those techniques. Students wishing to know more about that topic will need to consult more advanced texts on forest measurement (e.g. Philip 1994; Avery and Burkhart 2002; Husch et al. 2003; van Laar and Akça 2007) and the scientific literature.

It is impossible to teach forest measurement properly without a practical component to the course under the guidance of an experienced teacher. No book can substitute for that, so the reader of this book should expect only to be exposed to the principles of the discipline, rather than to become immediately competent in its practice.

Some terms used will be unfamiliar and a glossary has been included as Appendix 1. Terms in the glossary are shown in **bold** type when they are first encountered in the text. The metric system of weights and measures has been used throughout. To many North American readers in particular, this system will be unfamiliar and to them I apologise. I can only say with what relief, as a young forester in Australia in the 1970s, I welcomed the introduction of the metric system and could leave behind the horrors of the imperial system! A table of metric-imperial conversion factors has been included as Appendix 2. Younger readers, who have grown up with the metric system, will find there also some of the relationships between units in the imperial system; they can relish the realisation that they have not had to learn by heart such arcane facts as that there are 4,840 square yards in an acre.

There is little that can be done in **forest** measurement without using some mathematics. This book is designed so that a knowledge of no more than senior secondary-school-level mathematics is required; much will still be understood with a lower level of mathematical ability. There are many advanced techniques of forest measurement that require much higher-level mathematics; those are barely alluded to and certainly no detail is given.

Letters of the Greek alphabet are used commonly in mathematical formulae. A copy of the alphabet is included as Appendix 3 so that readers will be able to give names to the Greek letters as they are encountered. Many trigonometric concepts will be encountered also. The basics of trigonometry are summarised in Appendix 4.

1.2 What Measurements Are Considered?

It is impossible for any book to cover the whole range of things that might be measured on **trees** or forests. Pérez-Harguindeguy et al. (2013) published a handbook that summarised measurement techniques for what they considered a comprehensive range of plant characteristics that are measured commonly; they concentrated on those characteristics that plant scientists use in their research work. Surprisingly, they included little of the measurements or measurement techniques that are included in this book. Perhaps this illustrates that it is impossible to be completely comprehensive in any text on the practice of plant measurement.

The primary focus of this book is on measuring the trees in forests rather than the other biological organisms that occupy forests or the **environment**al characteristics of the **site**s on which forests grow. One principal concern is measurement of the amount of **wood** trees contain in their stems and the sizes of the logs that can be cut from them. It is this wood that is converted into **timber** products (lumber as it is called in America) for building and many other purposes or that is to be used for making paper. Wood in the form of logs cut from tree stems remains a valuable commercial product of forests; traditionally, university courses on forest measurement have concentrated on how it is measured.

A second concern of this book is with measurement of the weight of the various parts of trees, their leaves, branches, stems and roots. There are two reasons for this. Around the world, man uses about two billion dry tonnes of wood annually. Just over half of this is firewood, largely for domestic use, especially in Asia and Africa. That is to say, firewood is by far the biggest single use of wood by man. Firewood can be obtained from stem and branch wood of trees and, sometimes also, from large, woody roots and is usually measured by its weight.

The second reason for measuring tree weight is that about one quarter of the fresh weight of a tree (that is the weight of the tissue cut directly from a living tree) consists of the chemical element carbon (about one half is water). Of recent times, there has been great concern around the world about global warming. This has been attributed to the release into the atmosphere of greenhouse gases from burning fossil fuels, such as coal and oil, to produce energy for human use. Carbon dioxide is one such gas. Plants in general, not just trees, take in carbon dioxide through their leaves and convert it chemically to sugar; they then use the sugar as food for their growth and various life functions. This process is known as **photosynthesis**. It requires energy from sunlight and releases oxygen back into the atmosphere as a waste product: it is this 'waste' product of photosynthesis that we animals breathe. Because plants remove carbon dioxide from the atmosphere and store it in their tissues (albeit stored in the form of other carbon-containing chemical compounds), plants are now seen as tools in attempts to reduce carbon dioxide levels in the atmosphere. Thus, measurement of the amount of carbon that trees and forests around the world can store has assumed great importance over recent years.

There are of course many things other than tree stem wood volumes and tree weights that might be measured in forests. Information might be needed about the plants and other animals that live in forests, the soils on which they grow or the streams and rivers that run through them. These are important to understanding many of the other values that forests offer in matters such as conservation, recreation, the supply of clean water or the rehabilitation of degraded land. These values are being appreciated more and more today. However, their measurement and valuation are properly the subject of other books.

1.3   Scale of Measurement

This book is concerned with forest measurement at three scales. These range initially from individual trees then to **stand**s of trees (a stand is a more or less homogeneous group of trees in a forest in which an observer might stand and look about him or her) and finally to large forest areas. The book is structured to consider measurements at these successively larger scales.

Individual trees occupy only a few square metres of land, whilst whole forests may cover hundreds or thousands of hectares. Thus, the measurements that can be taken at the smallest of those scales are likely to be much more detailed than those taken over larger areas. Much of the measurement of forests at larger scales is concerned with making measurements at a small scale, then using mathematical techniques to bring those measurements up to a large scale. Much of the content of the book is concerned with those techniques of scaling up.

Perhaps surprisingly, it is quite possible to take tree measurements using very simple equipment like hand-held tapes. These simple devices have been the mainstay of forest measurement over the last century or so. However, their use is labour intensive and requires that field measurement crews travel around the forest area being considered and take their tree measurements directly in the forest.

There is now an increasing desire to use far more sophisticated equipment in forest measurement. Of course, a computer is used generally to assist both in storing the data collected from the forest and to do the arithmetical computations needed to convert those raw data into useful information about the forest. But highly sophisticated measuring devices, ranging from **digital** cameras used on the ground to satellite images of the forest made from space, are now being adapted for use in measuring trees and forests. Not only is this equipment likely to be labour saving but it will allow much larger areas of forest to be measured in far more detail than was possible in the past. The final chapter of this book is devoted to a description of some of this equipment and its use from small- to broad-scale measurement of trees and forests.

Chapter 2
Measurements

Abstract When measuring anything, the accuracy required of the measurement, the possibility of bias in it and its precision must all be considered. Accuracy concerns how closely to its true value one can measure something. It is determined by the quality of the measuring equipment and is expressed as a measurement to the nearest part of some measurement unit, for example, to the nearest metre. If something is very difficult to measure, the measurement method may consistently give an over- or underestimate of the size of the thing; the measurement is then said to contain bias. As well, if the thing is difficult to measure, we may not feel very confident that the measurement we have taken reflects truly the actual size of the thing; the degree of confidence we have in the measurement is known as its precision. Judgement is required by the measurer to assess what level of accuracy is required of a measurement and whether or not the degree of bias in it and its precision render the measurement suitable for the purposes for which it is required. This chapter discusses these issues.

2.1 Measuring Things

Measurement of things is a fundamental part of any scientifically based discipline. Some things are simple to measure, like the length of a piece of string or the time it takes a pedestrian to cross the road. Other things are very difficult to measure, like the size of an atom or the distance to Jupiter. Some things cannot be measured directly at all, like the volume of wood that might be harvested from a large forest area of thousands of hectares; there are simply too many trees in such a forest to measure them all and, as will be seen in Chaps. 5 and 6, it is quite difficult to measure the harvestable wood volume in even just one tree.

When something is difficult to measure, or cannot be measured directly at all, methods of measurement are used to approximate or estimate it. These methods often involve measuring parts of the thing, parts that can be relatively easily measured. Then, more or less complicated mathematical procedures are used to convert the measurements of the parts to make an estimate of the size of the whole thing.

© Springer International Publishing Switzerland 2015
P.W. West, *Tree and Forest Measurement*, 3rd edition,
DOI 10.1007/978-3-319-14708-6_2

Indeed, this book is concerned both with how parts of things in forests are measured, simple parts like the circumference of the stem or the **height** of a tree, and how those simple measurements are used to estimate a more difficult thing, like the harvestable wood volume in its entire stem.

Whether a simple or very complex thing is being measured, there are three things about its measurement with which we should be concerned. These are the **accuracy** of the measurement, whether or not there is **bias** in it and what is its **precision**. The rest of this chapter will be concerned with these three issues in the context of measurement of trees and forests.

2.2   Accuracy

Accuracy is defined formally as 'the difference between a measurement or estimate of something and its true value'. In simple terms, it can be thought of as how closely one is able to measure or estimate something given the measuring equipment or estimation method available. Accuracy is expressed by saying that a measurement or estimate has been made to the nearest part of some unit of measurement, for example, to the nearest 1/10th of a metre, to the nearest hectare or to the nearest microsecond, depending on what type of thing is being measured.

Suppose it was desired to measure something quite simple, like the length of the side of a field of which the true length was exactly 100 m. There are a variety of methods that could be used to do that. The simplest might be to simply pace the distance out yourself, having calibrated your paces by measuring their length along a tape measure. However, a result from pacing would not be expected to be very accurate because a person is unable to keep each of his or her paces exactly the same length. Pacing would probably give a result for the length of the side of the field somewhere in the range of about 95–105 m. That is, we could then say that measuring distances of around 100 m by pacing was accurate only to the nearest 5 m.

A second method might be to use a measuring tape. Such tapes are often 30–100 m long, made of fibreglass or another material that is not likely to stretch, and are usually calibrated in 1 cm units. Some care is needed with their use; they must be laid carefully along the ground and pulled tight to ensure that dips, hollows and irregularities in the ground surface influence the result as little as possible. However, even taking all due care with a tape like this, it would probably give a result for the length of the side of the field somewhere in the range 99.9–100.1 m. That is, we would say the tape was accurate to the nearest 1/10th of a metre.

A third method might involve a modern **laser** distance measuring device, such as that used today by professional surveyors. Lasers are becoming very important for many types of measurement, not only in **forestry**; their use in forestry is discussed further in Chaps. 4, 5 and 13.

Laser is an acronym for 'light amplification by stimulated emission of radiation'. Laser light involves an intense, narrow beam of light of a single colour that can be directed very precisely. The distance from an instrument to a solid object is determined by measuring the time it takes a pulse of laser light to be reflected from the object back to the instrument. These instruments contain very accurate clocks capable of measuring the extremely short periods of time involved, given that light travels at about 300 million metres/second. A laser distance measuring device might be capable of measuring a distance of about 100 m with an accuracy at least to the nearest 1/1,000th of a metre, that is, to the nearest millimetre.

The size of the thing being measured will immediately set some criterion for the accuracy required of the measurement. If one wishes to measure the sizes of atoms that are of the order of 1 angstrom unit in diameter (an angstrom unit is one 100 millionth of a centimetre and was named for Anders Ångström, a Swedish physicist of the mid-nineteenth century), complex laboratory equipment will be required capable of taking measurements with an accuracy of fractions of an angstrom unit. If one wishes to measure the distance to Jupiter, which orbits the sun at an average distance of about 778 million km, a measurement method accurate to the nearest few tens of thousands of kilometres is probably what is required. However, the accuracy required ultimately of a measurement or estimate of something depends on the purpose for which the result is required. In turn, this will determine the sophistication of the equipment or estimation method required to achieve the desired accuracy.

Returning to the simple example of measurement of the length of the sides of a field, if it was desired to determine its area roughly to work out how many bags of fertiliser were needed to cover it, the accuracy of measurement got from pacing out the sides would probably be adequate. On the other hand, if a professional surveyor wished to measure the field to establish the title to the property, a laser measuring device would probably be preferred to achieve the accuracy required by the legal system.

2.3 Bias

Bias is defined as 'the difference between the average of a set of repeated measurements or estimates of something and its true value'. In essence, if something is difficult to measure, it may not matter how many times we attempt to take the measurement nor how many different types of measurement equipment we use, we may nearly always get the wrong answer. By 'the wrong answer' is meant that the results of the many attempts at measurement will be consistently larger or smaller than the true value of whatever it is that is being measured. That is not to say there has been a mistake in taking the measurement, but simply that what is being measured or estimated is too difficult and we are unable to take the measurement exactly. When this occurs, the measurement or estimation method is said to be biased.

By the same token, it would be said that the measurement or estimation method is unbiased if the average of the many measurement attempts differed negligibly from the true value. How small would the difference have to be considered negligible? Obviously, some limit is set by the accuracy of the measurement method; we simply cannot detect differences smaller than the accuracy. Apart from that, the degree of bias that will be considered acceptable will be determined entirely by the purposes for which the result of the measurement is to be used; this issue is discussed further in Sect. 2.5.

To illustrate what is meant by bias, consider the problems involved in measuring the weight of the fine roots of a tree. Fine roots are the small (less than about 2 mm diameter), live roots at the extremities of the root system. Biologically, they are extremely important, because they take up the water and **nutrients** from the soil that the tree needs to survive and grow. Because of their importance, forest scientists need to measure them. The most appropriate way devised so far to do so is to excavate them from the soil. Obviously, this is a major task since they will be scattered throughout a large volume of soil, extending perhaps 2–3 m or more away from the stem of a large tree and to a depth of 1–2 m. As well, so small and numerous are fine roots, it is very difficult to find all of them as one sorts laboriously through such a large volume of soil. Furthermore, in any patch of forest, it is difficult to know if an excavated fine root belongs to the particular tree one is dealing with or if it belongs to another nearby tree or even to an **understorey** plant. So difficult are fine roots to find and measure, it is perhaps inevitable that any attempt to do so is doomed to get the 'wrong answer', that is, to be a biased measurement method. Most probably, the answer will be an underestimate of the true amount because it is so difficult to find all the fine roots. There are various other methods used to measure fine roots (Sect. 7.2.3), all of them probably subject to bias because of the difficulties associated with their measurement.

2.4 Precision

Precision is defined as 'the variation in a set of repeated measurements or estimates of something'. The variation arises because of the limitations of the people taking the measurements and the limitations in the measurement or estimation technique when it is used at different times and under varying circumstances.

Following the example in Sect. 2.3, if a number of different people set out to measure the weight of the fine roots of a tree, it is inevitable that each would get a somewhat different result. So difficult are fine roots to measure that individuals will vary in how many they manage to find in a large, excavated soil volume.

Precision is measured by the amount of variation in the results of a repeated set of measurements of the same thing. The range of values in the set of estimates is one measure of precision. Another measure, called **variance**, is the measure used most commonly. Variance is a concept that derives from **mathematical statistics**, a branch of mathematics much concerned with the problems that variation between natural things causes us in understanding how nature works. Variance and its use as a measure of precision will be discussed more fully in Chap. 9.

Suppose the precision of a measurement or estimation technique is low. That is, a rather wide range of results would be obtained if the technique was used by different people or at different times to measure the same thing. In practice, we usually only measure or estimate something once using whatever measurement equipment or estimation technique is available to us. If we know that the precision of the technique we are using is low, we would feel rather unsure about the extent to which we could rely on that single result we had obtained. In turn, we would not be very confident that we could draw worthwhile conclusions about whatever it was that was being measured. That is why precision is important in measurement; it is a measure of how confident we feel that a measurement we have taken of something truly represents its real size. If the precision of a measurement is high, we will feel confident that we can use the information to draw reliable conclusions. If it is low, we will feel much less confidence in our conclusions.

2.5   Bias, Precision and the Value of Measurements

It is important to understand how bias and precision interact. This can be illustrated through an analogy used in various texts (Shiver and Borders 1996; Avery and Burkhart 2002), where a marksman is shooting at a target. In effect, the marksman is attempting to use a bullet to 'measure' the position of the bullseye of the target.

Figure 2.1 describes the analogy. The best possible result for the marksman is illustrated in Fig. 2.1a. The average position of all the shots is right on the bullseye; that is, the average of the repeated attempts to measure the position of the bullseye does not differ appreciably from its true position, so it can be said to have been an unbiased measurement technique. As well, because the shots cluster closely around the bullseye, it can be said they measure its position with a high degree of certainty and so they represent measurements made with a high degree of precision.

In the case of Fig. 2.1b, the shots still cluster closely around one point, so they represent measurements made with a high degree of precision. However, their average position is some distance from the bullseye, so they represent a measurement technique in which there is bias. In this analogy, the bias might have arisen because the 'instrument' being used (the gun) is not calibrated correctly, by having its sights set poorly. Or perhaps, unknown to the marksman, there was a wind blowing that pushed all the shots to the left.

Figure 2.1c and d both show cases where the marksman has produced a wide spread of shots that represent measurements made with a low degree of precision. In Fig. 2.1c, despite their wide spread, the average position of the shots is still right on the bullseye, so they represent measurements made without bias. This might happen to a marksman on a day when the wind varies unpredictably, so that his or her shots are spread. Figure 2.1d represents the worst possible result for the marksman. Not only are the shots widespread, but also their average position is a long way from the

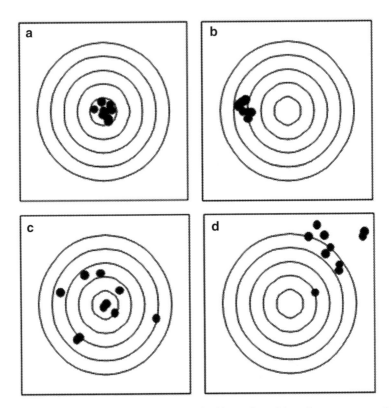

Fig. 2.1 *Bullet holes* in a target as an analogy for bias and precision of measurements. (**a**) An unbiased, precise result; (**b**) a biased, precise result; (**c**) an unbiased, imprecise result; and (**d**) a biased, imprecise result

bullseye. This might happen if the sights of the gun are not set correctly and if there are unpredictable wind variations.

The important question then is whether or not a biased or imprecise measurement is still useful. Usually, it is better to have some measurement of something than no measure at all. What is difficult to judge is whether or not a biased but precise result (Fig. 2.1b) is more useful than an unbiased but imprecise result (Fig. 2.1c). Even more difficult to judge is if a biased and imprecise result (Fig. 2.1d) is better than no result at all. There are really no rules available to make these decisions. It becomes a matter of judgement for the person using the results to decide whether or not they are adequate for the purposes for which they are needed.

As discussion of various measurement techniques continues throughout this book, reference will be made to the accuracy, bias and precision involved with them.

Chapter 3
Stem Diameter

Abstract The girth or diameter of the stem of a tree is arguably the most useful thing measured on trees in forest science. It correlates closely with many other things that are much more difficult to measure, such as the amount of wood in the stem of the tree, its total biomass (weight) or biomass of its parts (leaves, branches, stem, roots) or its competitive position in the forest. A worldwide convention is to measure tree stem diameter at 'breast height', 1.3 m above ground (a slightly different height is used in a few countries). Most commonly, stem diameter is measured using a girth tape. This chapter discusses how and why tree stem diameters are measured both at breast height and at other heights up the stem if needed. Problems that can arise with these diameter measurements are discussed.

3.1 Basis of Diameter Measurement

The simplest, most common and, arguably, the most useful thing measured on trees in forestry is the diameter of their stems. Amongst other things, tree stem diameter:

- Often correlates closely with other things that are more difficult to measure, like the wood volume in the stem of a tree or the weight (or **biomass**, as it is called) of the tree
- May reflect the monetary worth of the tree, given that larger trees produce logs of larger sizes from which more valuable timber can be cut and so are more valuable commercially
- May reflect the competitive position of a tree in a stand and, hence, how well it is likely to grow in relation to the other trees

Stem diameter declines progressively from the base of the stem as the tree tapers. So, a standard convention has been adopted in forestry to make a basic measurement of tree stem diameter at **breast height**. This is defined as being 1.3 or 1.4 m vertically above ground from the base of the tree. The height used varies in different countries (and is actually defined in imperial units as 4 ft 6 in. in America); the difference is generally ignored when results from different countries are compared. If the tree is growing on sloping ground, breast height is measured from the highest

© Springer International Publishing Switzerland 2015
P.W. West, *Tree and Forest Measurement*, 3rd edition,
DOI 10.1007/978-3-319-14708-6_3

ground level at the base of the tree. Loose litter and debris at the base of the tree should be brushed aside before making the measurement of breast height. Of course, stem diameters may be measured also at heights along the stem other than breast height; reasons for doing so are discussed in Sect. 5.3.4.

If a tree is very young, it may not have grown tall enough to have reached breast height. If it is desired to measure its stem diameter, obviously it must be done at a lower height, at least until the tree is taller than breast height; commonly, heights of 0.1 or 0.3 m above ground are used in these circumstances. The need to do this is increasing in forestry as some products are now harvested from very young forests. For example, **plantation** forests are being grown for only 3–5 years to produce wood for **bioenergy** production (i.e. to fuel boilers or to be converted by fermentation to ethanol for motor vehicle fuel). However, forestry has not yet adopted any particular convention as to the height to be used for stem diameter measurement of small trees.

The rest of this chapter discusses how stem diameters are measured and the difficulties encountered in doing so.

3.2 Stem Cross-Sectional Shape

By referring to diameter, it is being implied that stems are circular in cross section; this is never exactly so. Certainly they are approximately circular because the principal function of the stem is to act as a pole and support the **crown** (the leaves and branches of a tree) high in the air so that the tree can dominate the other vegetation that occurs on a site. Engineering theory suggests that a circular pole will be stronger than poles of the same size but of other shapes; thus, it can be argued that evolution has favoured the development of tree stems of the most efficient shape to perform their function.

However, all stems have some irregularities in their cross-sectional shape simply because trees are biological organisms and nature rarely provides theoretical perfection. Those irregularities are generally exaggerated at points where branches have protruded from the stem or where damage has occurred through things like fire, disease or insect attack (Fig. 3.4). As well, most stems show some eccentricity in their shape so that they are wider in one direction than another. This is most likely a response to wind; the longest axis of the eccentric shape will correspond to the prevailing wind direction and will give the stem more strength to resist those winds. In fact, the density of stem wood has also been found to be greater along the axis of the prevailing wind direction, an effect that also increases the strength of the stem in that direction (Robertson 1991).

Particularly in tropical **rainforest**s, large trees may have extensive flutes or buttresses protruding from their bases (Fig. 3.1). These may extend to several metres above ground. Just like buttresses used in buildings, tree buttresses are believed to give additional structural support to the tree but may have other effects on the forest **ecosystem** (Pandey et al. 2011).

Fig. 3.1 Buttressing on the
lower stem of a large tree in
subtropical rainforest in
northern New South Wales,
Australia. This stem is over
3 m wide at its base. The
buttressing continues up the
stem for more than 5 m
(Photo—P.W. West)

Apart from these common irregularities in the cross-sectional shape of tree
stems, extraordinary variations in shape occur also. Generally, these are a result
of unusual environmental circumstances, where trees lean against one another or
some other solid object, grow on steep slopes or have odd branching. In his unusual
and entertaining book, Dr. Claus Mattheck has illustrated some of the extraordinary
shapes that trees have been found adopting in nature (Mattheck 1991). These
unusual cases are sufficiently rare that they need not be of concern for normal
forestry circumstances.

Given all this discussion, it is clear that tree stems are generally not exactly
circular in cross section. This means that stem diameter will usually be a biased
measure of the true size of the stem. Studies have suggested this may lead to a bias
of around 2–3% in determination of stem cross-sectional area (Gregoire et al. 1990;
Pulkkinen 2012) or growth in stem cross-sectional area (Biging and Wensel 1988).
This may be a sufficiently large bias to be of concern in detailed research studies.
However, for many forestry purposes, it is sufficiently small that it is generally
ignored, and tree stems are treated as being truly circular in cross section.

3.3   Measuring Stem Diameter

The most common way to determine the diameter of a stem is to measure its girth with a simple tape measure, known as a diameter tape. Diameter tapes are made of steel or fibreglass, for strength and to prevent stretching. They are calibrated in units of the mathematical constant pi (π), which is the ratio of the circumference of any circle to its diameter and has a value of approximately 3.142. That is, a unit shown as 1 cm long on a diameter tape is 3.142 cm long; when the tape is wrapped around the girth of a tree, the corresponding diameter can be read directly from the tape.

To use a diameter tape correctly, it should be wrapped firmly around the stem, perpendicular to its axis (Fig. 3.2). Any loose bark should be brushed gently off the

Fig. 3.2 Measuring the diameter at breast height over bark of a spotted gum (*Corymbia citriodora* var. *variegata*) tree in a forest plantation in northern New South Wales, Australia. This species sheds loose bark naturally so there was none to remove here before measurement. Note how the diameter tape is being held firmly and horizontally around the stem (Photo—P.W. West)

stem before making the measurement. Where a tree is to be measured repeatedly to determine its growth rate, say, at intervals of a year or so, paint or other marking material may be used to mark the point where the diameter is measured to ensure the same position is measured on each occasion.

Diameter tapes are usually calibrated at intervals of 0.1 cm diameter (i.e. the calibration marks are 3.142 mm apart), and tree measurements are usually recorded to an accuracy of the nearest 0.1 cm (i.e. to the nearest millimetre). Years of experience of forest scientists have shown that this accuracy is adequate generally for forestry purposes.

A second instrument used commonly to measure diameter is a caliper (Fig. 3.3). Calipers are particularly useful when measuring trees of small diameter (say, less than about 5 cm), when the stiffness of a diameter tape can make it difficult to wrap the tape around the stem. However, calipers are used also to measure trees of larger diameter, the size of the calipers being chosen to suit the size of the trees being measured. Calipers are often quicker to use than diameter tapes. However, they measure stems only across one diameter of their cross section, whereas a diameter tape measures the average diameter corresponding to the girth of the tree. To allow for this, it is usual when using calipers to take two diameter measurements at right angles to each other. The square root of the product of the two diameters is then used as the measure of stem diameter; by calculating stem diameter this way, it is being allowed that the stem cross section may be shaped as an ellipse, rather than being circular.

Much less commonly than diameter tapes or calipers, other instruments are used to measure tree diameters, such as Biltmore sticks or the Wheeler pentaprism. They are used less commonly and are described in some other books on forest measurement (Philip 1994; Avery and Burkhart 2002; van Laar and Akça 2007); they will not be considered further here. There are available also optical instruments with which stem diameters can be measured high up on the tree stem. These will be discussed in more detail in Sect. 5.3.4.

Fig. 3.3 Calipers used to measure tree stem diameter. They are available in various sizes to deal with tree stems of different diameters (Image provided by courtesy of Forestry Tools Australia, http://www.forestrytools.com.au)

3.4   Stem Irregularities

A common problem in diameter measurement results from the lumps and bumps that may occur anywhere along a tree stem. These are especially common where branches protrude and may persist for some years even after a branch has died and dropped off. When such an irregularity occurs where a diameter measurement is to be made, two measurements are usually taken at points equidistant above and below the point. The average of the two measurements is then used as the measurement of stem diameter at the required point. It is left to the judgement of the measurer to assess if such an irregularity is sufficiently large to warrant measuring diameter in this fashion (Fig. 3.4).

Also common is the occurrence of trees with forks in the stem, beyond which the tree has grown with two or even more stems. There are many tree species also that have multiple stems arising from ground level. The convention used to deal with these cases is to treat the multiple stems as separate trees whenever the fork occurs below breast height.

Where buttresses occur (Fig. 3.1), the stem is so irregular in shape that it is obviously quite impossible to define its diameter simply. Where stem distortion due

Fig. 3.4 An example of a badly deformed tree stem, in this case as a result of attack by a wood-boring insect and through branch shed. The two *arrows* indicate where stem diameter measurements might be made, above and below the deformation, and then averaged to give an estimate of the stem diameter halfway along the defect. In this case, the stem irregularity was so marked that the two diameter measurements would be taken nearly 1 m apart (Photo—P.W. West)

to buttressing occurs above breast height, a convention has been suggested for general use to measure stem diameter at a height of 30 cm above the point where the effect of the buttressing has disappeared and where the stem has become approximately circular in cross section (Branthomme et al. 2004). This height may vary from tree to tree, depending on how the buttressing occurs. It may be several metres above ground, when a ladder would be needed to reach the required height.

Such measurements may be applied to all the purposes for which breast height diameters are used normally and that will be discussed in due course in this book. However, they are obviously no longer directly comparable with measurements made at breast height, the world forestry standard.

Ngomanda et al. (2012) undertook research that suggested another approach to this problem. They measured 102 trees with buttresses of 31 different species in the tropical moist forests of Gabon, Africa. Their stem diameters immediately above the buttress varied over the range 26–95 cm. The convoluted cross-sectional area at breast height of each stem was measured in detail. This area was converted to an equivalent diameter (D, cm) at breast height by assuming the stem was circular rather than convoluted. They also measured the girth of the stem at breast height (G, m), with a tape held firmly around the stem over the buttress flutes, and the girth (G',m) at the point above the buttress, where the stem started to adopt a convex shape. They found they could estimate quite reliably the equivalent diameter at breast height from either of these girth measurements as

$$D = 5.71 + 23.11G \qquad (3.1)$$

or

$$D = 34.73G' \qquad (3.2)$$

Further research is needed to examine this approach with other tree species in other parts of the world. However, it holds promise as a method to determine diameters at breast height of buttressed trees that are comparable with breast height diameters of trees without buttresses.

3.5 Bark Thickness

Forestry is concerned usually with the wood in tree stems, because that is the part of the tree that is sold most commonly. Bark may be sold also, perhaps as mulching material for gardens, or it can even be burnt as a biofuel. However, it is generally a much less valuable product than wood so it is usually the wood it is desired to measure.

Between different tree species, bark varies greatly in thickness and texture, from extremely rough to quite smooth. It can be several centimetres thick, so a

Fig. 3.5 A bark gauge that measures bark thicknesses over the range 0–5 cm. When its left-hand end is pressed against the tree, a borer protrudes through the bark until the wood below is encountered. The thickness of the bark is then read from the scale (Image provided by courtesy of Forestry Tools Australia, http://www.forestrytools.com.au)

measurement of stem diameter made over the bark can be appreciably greater than the diameter of the wood below.

Bark thickness of standing trees can be measured with a bark gauge (Fig. 3.5). This instrument consists of a shaft with a sharp point that is pushed through the bark until the resistance of the underlying wood is felt. The sleeve around the shaft is then shifted to the surface of the bark and the bark thickness read from the calibrated shaft. Some practice is needed to get a 'feel' for when the point of the gauge has reached the wood. Usually, at least two measurements, at right angles around the stem, would be made of bark thickness and their average used as the measure of bark thickness.

Measuring bark thickness can be quite tedious. So, wherever possible, measurements of stem diameter over bark are preferred. As shall be seen below, over bark diameter measurements are quite adequate for many of the purposes for which stem diameter measurements are used in forestry. However, there are times when it is essential that under bark diameters be determined and so bark thickness must be measured.

Chapter 4
Tree Height

Abstract The height of a tree is important in calculating the amount of wood its stem contains, in determining the competitive position of a tree within the forest and in assessing the productive capacity of the site on which the tree is growing. In forestry, it is defined as the vertical distance from ground level to the highest green point on the tree. This chapter describes basic methods of tree height measurement. Height may be measured directly by holding a pole alongside the tree. For trees taller than about 12–15 m, a trigonometrically or geometrically based method may be used. The latter requires no more than a 3–5 m long stick and a ruler. More sophisticated electronic instruments are available today to speed the height measurement of taller trees.

4.1 Basis of Height Measurement

The height of trees is important to forestry particularly because:

- The length of the stem is important as part of the calculation of the total amount of wood contained within it.
- It may reflect the competitive position of a tree in a stand and, hence, how well it is likely to grow in relation to the other trees.
- The height of the tallest trees in the forest is the basis of one of the most important measures used in forestry to assess **site productive capacity**. This is a measure used to assess how rapidly trees will grow on a site; it will be discussed further in Sect. 8.7.

In forestry, tree height is defined as the vertical distance from ground level to the highest green point on the tree (a point that will be referred to here as the 'tip' of the tree). It might seem odd that tree height is not defined in terms of stem length (since it is usually the wood-containing stem of the tree with which forestry is most concerned) or as the height to the top of the stem itself. However, near the tips of trees, it is usually difficult to define exactly what constitutes the stem because of the proliferation of small branches there. Even if the main stem can be seen clearly near

© Springer International Publishing Switzerland 2015
P.W. West, *Tree and Forest Measurement*, 3rd edition,
DOI 10.1007/978-3-319-14708-6_4

the tip, it is often very difficult to see exactly where it stops. This is particularly so when viewing a tall tree with a dense crown from the ground.

Whilst the highest green point of a tree is much easier to identify than its stem length, care must be taken to ensure that the tree is viewed from sufficiently far away so that its tip can be seen clearly. Even then, in dense forest it is often difficult to see the tip amongst the crowns of other trees; care must be taken to ensure the tip one can see is indeed that of the tree being measured.

Even if the tree is leaning, its height is still defined in forestry as the height to the highest green point, rather than by its stem length. Most trees, in most forest circumstances, stand just about vertically; if they do lean a little, perhaps in response to strong prevailing winds or a marked ground slope, the lean is usually no more than a few degrees. For general forestry purposes, it is sufficiently rare to encounter trees leaning sufficiently that special consideration has to be given to how their height should be measured; the lean would have to exceed about 7–8° before it would be sufficient to affect appreciably the result of a tree height measurement, given the accuracy with which we normally measure tree heights. Heights of leaning trees will not be considered further here.

Direct, trigonometric and geometric methods are used to measure tree heights. Each of these will be discussed below.

4.2   Height by the Direct Method

Direct height measurement involves simply holding a vertical measuring pole directly alongside the tree stem. Devices with a telescoping set of pole segments can be purchased readily. These are able to measure tree heights to about 8 m.

Lightweight aluminium or fibreglass poles of a constant length (1.5–2 m), that slot into each other at their ends, are available also. As many as necessary of these may be slotted together progressively and the whole lot raised until the tip of the tree is reached. The number of poles used is counted and any leftover length at the base of the tree is measured with a tape. These are effective to heights of about 12–15 m, beyond which the poles become too heavy or unwieldy to hold.

When using these devices, care must be taken to ensure the pole is raised to coincide exactly with the tip of the tree. This requires a team of two, one to hold the measuring pole and the other to sight when the tip of a tree is reached. In windy weather, swaying of the tree tops can make this sighting more difficult.

With careful sighting of the tree tip, these devices should allow height measurements to an accuracy of about the nearest 0.1 m. For trees taller than about 12–15 m, it is necessary to use the trigonometric or the geometric methods as discussed below.

4.3   Height by the Trigonometric Method

Figure 4.1 illustrates the principle involved in measuring tree height by the trigonometric method. A vertical tree of height $h_T = AC$ is standing on flat ground. An
observer is standing a measured distance $d = GC$ away from the tree and measures,
at eye level O with some viewing device, the angles from the horizontal to the tip of
the tree, a_T, and to the base of the tree, a_B. Angles measured above the horizontal
should have a positive value, whilst those below the horizontal should be negative;
in the case of Fig. 4.1, a_T is positive and a_B is negative.

Using straightforward geometry and trigonometry, the height of the tree can be
calculated from these measurements as

$$h_T = d[\tan{(a_T)} + \tan{(-a_B)}] \tag{4.1}$$

where 'tan' is the trigonometric expression for the tangent of the angle. Appendix 4
summarises some basic trigonometric principles.

As an example, suppose the observer was standing 21 m away from the tree and
measured the angle to the tip as 48° and the angle to the base as −7°. Then, the
height of the tree would be calculated as $h_T = 21 \times [\tan(48) + \tan(7)] = 21 \times
[1.1106 + 0.1228] = 25.9$ m. Scientific calculators and computer programs provide
the required trigonometric functions.

In dense forest it can often be difficult for the observer to see the tip of the tree.
He or she needs to move around the tree and adjust the distance from which it is
being viewed to make sure that the tip can be seen clearly. These problems are
exacerbated if the wind is blowing the tips about. If the day is too windy, it simply
becomes impractical to undertake height measurements.

A tape may be used to measure the distance from the observer to the centre of the
base of the tree. The angles may be measured with a hand-held clinometer (readily
available from forestry suppliers) or, more precisely, with a theodolite. Theodolites

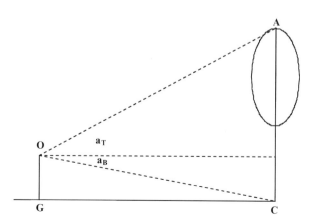

Fig. 4.1 Principle of tree
height measurement using
the trigonometric method

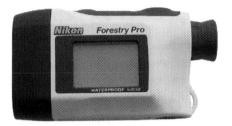

Fig. 4.2 A modern, hand-held instrument that uses the geometric method to measure tree height. In this case, distance to the tree is measured using a laser and a clinometer is built in. Immediately the measurement is completed, the device calculates the height electronically and displays the result on its screen (Image provided by courtesy of Forestry Tools Australia, http://www.forestrytools.com.au)

are far slower to use and would only be countenanced if a very precise height measurement was required.

For routine tree height measurements, convenient electronic instruments are available today. These combine a clinometer with a distance measuring device. Some use the time of travel of sound waves to measure the distance, whilst the most recent use a laser (Fig. 4.2). In both cases, a target is pushed into the stem of the tree to reflect back to the instrument the sound wave or laser light. Because the velocity of sound varies appreciably with air temperature, the instruments that use sound need to be calibrated regularly throughout the day as temperature changes. Once distance has been measured, the instrument is aimed at the base and tip of the tree and the inbuilt clinometer measures the required angles. The tree height is then calculated electronically by the device and displayed to the user.

Heights measured by the trigonometric method are often reported to an accuracy of the nearest 0.1 m. However, given the difficulties involved in sighting to the tips of tall trees, this is probably optimistic. In the example given above, a measurement error as small as +0.5° in the angle to the tip of the tree would result in a height estimate of 26.3 m, rather than the 25.9 m given in the example. In practice, an accuracy of no better than to the nearest 0.5 m might be a more realistic assessment for tree height measurements using hand-held devices.

Often the land surface on which the tree is positioned is sloping, rather than flat as in Fig. 4.1. To allow for this, the observer needs to measure also the angle of the slope, a_S. This may be positive or negative, depending on whether the observer is positioned down- or upslope, respectively, from the tree. The slope angle may be measured with a clinometer as the angle from the horizontal to a point on the stem at a height equal to the observer's eye level. The distance from the tree to the observer is then measured along the slope. Say the slope distance is s, then the horizontal distance to the base of the tree, d, can be calculated as

$$d = s \cos (a_S), \tag{4.2}$$

where 'cos' is the trigonometric expression for the cosine of an angle. Suppose the slope angle was a downslope of $-10°$ and the slope distance measured was 21.3 m, then the horizontal distance to the tree would be calculated as $d = 21.3 \times \cos(-10) = 21.3 \times 0.9848 = 21.0$ m. The angle to the tip and base of the tree would be measured as described before, and this horizontal distance would then be used in (4.1) to calculate the tree height.

On steeply sloping ground and where the observer is standing downslope of the tree, the angle measured to the base of the tree, a_B, may be positive, rather than negative as in Fig. 4.1. This does not affect the computation of height in any way, and (4.1) and (4.2) remain appropriate to calculate the height of the tree.

The sonic or laser measuring devices described above adjust automatically for ground slope by measuring the angle up or down to the reflector on the tree that is always positioned at a standard height above ground.

4.4   Height by the Geometric Method

Figure 4.3 illustrates the principle involved in measuring tree height by the geometric method.

A vertical tree of height $h_T = AC$ is standing on flat ground. A straight stick of known length $h_S = BC$ is positioned vertically at the base of the tree; such a stick would commonly be about 3–5 m long. An observer is standing a convenient distance away from the tree, with his or her eye at O. The observer holds a graduated ruler DF, positioned so that the line of sight OC to the base of the tree is coincident with the zero mark of the ruler. Without moving his or her head up or down, the observer reads from the ruler the distance $l_S = FE$ that coincides with the line of sight OB to the top of the stick against the tree. He or she reads also from the ruler the distance $l_T = DF$ that coincides with the line of sight OA to the tip of the tree. Using straightforward geometry, the height of the tree can then be calculated from these measurements as

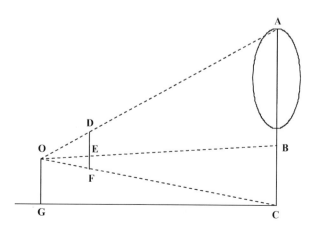

Fig. 4.3 Principle of tree height measurement using the geometric method

$$h_T = h_S(l_T/l_S) \tag{4.3}$$

As an example, suppose the length of the stick standing against the tree was 5.0 m and the observer measured l_T as 41.4 cm and l_S as 8.0 cm. Then, the height of the tree would be calculated as $h_T = 5.0 \times 41.4/8.0 = 25.9$ m. Ground slope does not affect the geometry of this method.

A number of different devices are available that use this principle. Often, the ruler is graduated in such a way that the computations in (4.3) are done implicitly, so that the tree height can be read directly from their scale. These devices are known generally as **hypsometer**s.

The difficulties of measurement that apply with the trigonometric method apply also with the geometric method. One advantage of the geometric method is that neither the distance from the observer to the tree nor ground slope need to be measured. A second advantage is that the equipment required is very simple (a stick of known length and a ruler only are required). Perhaps the disadvantage of the geometric method is that it is quite difficult physically for the observer to hold the ruler steady and, at the same time, keep in view all that needs to be sighted. However, with care, the accuracy of measurement of tree height using the geometric method should be about to the nearest 0.5 m, the same as that with the trigonometric method.

Chapter 5
Stem Volume

Abstract The volume of wood in tree stems is important to know commercially and for determining biomass. This chapter describes how direct measurements are made of tree stems and of logs cut from them. 'Exact' volumes of felled stems can be measured using water displacement or laser scanning and of standing trees using digital photography or laser scanning; none of these methods is entirely satisfactory for field use. Direct field measurement of tree stems or logs involves measurement of stem diameter at many points along them. The volume of each section is determined and they are summed to give the total volume; the more measurements taken, the more precise are such volume estimates. On felled stems, these measurements are made using a measuring tape and a diameter tape. High-quality ocular instruments are used to take these measurements from the ground on standing trees. The 'centroid' method can be used to reduce the measurement required for standing trees, usually with adequate precision. It requires measurement only of the diameter at breast height over and under bark of a tree, its total height and the measurement with an ocular instrument of a single diameter high up on the stem.

5.1 Reasons for Volume Measurement

The volume of wood contained in the stem of a tree is one of the most important measurements made in forestry, because:

- Wood is the principal commercial product of forests.
- The stem contains a very large proportion of the biomass of a tree.

Not only is the total volume of the wood in the stem of a tree of interest, but also the volumes of individual lengths cut from the stem, that is, of logs. Logs of different sizes, both in diameter and length, have different uses. Usually logs of larger diameter are required for conversion to solid wood products (i.e. sawn in a sawmill to make all sorts of building materials and many other products). Generally, these larger logs attract a much higher price per unit volume of wood than do smaller logs that may be suited only to chipping for papermaking. As well as the size of a log, its quality is important also. Factors such as its straightness, the

© Springer International Publishing Switzerland 2015
P.W. West, *Tree and Forest Measurement*, 3rd edition,
DOI 10.1007/978-3-319-14708-6_5

presence and size of branch knots and the presence or absence of any decayed wood within the log can all be important in determining its value.

Any one tree may contain a wide range of different log sizes. Larger logs are cut from nearer the base of the stem and smaller ones from further up. There will usually be parts of the stem near the tip of the tree that are too small to use for any product; these are usually left as waste on the ground when a forest is logged.

These days, logs are often sold by weight, rather than by volume, because it is easier to allow trucks carrying logs to a mill to pass over a weighbridge than to measure the volume of the logs on it. However, the logs on any one truck will usually have been sorted at the time of felling into logs of a particular size class and, hence, value. Implicitly, this means that logs have been sorted on the basis of their volumes and their conversion to weight is made simply on the basis of **wood density**; an example of this is given by Hultnas et al. (2013). In essence, then, volume remains the important variable for the characterisation of log size.

Forest science is often concerned with the production of biomass by trees; scientists who study the factors that affect tree growth behaviour often need to know how much biomass is contained in various parts of the tree (leaves, branches, bark, stem, coarse roots and fine roots). Chapter 7 will discuss the measurement of tree biomass. Since the stem contains a large proportion of the biomass of a tree, a proportion that increases with age as the stem continues to grow larger and larger, its measurement is very important. As will be seen in Chap. 7, stem biomass is often derived from stem volume by multiplying the volume by wood density. Thus, the issues discussed here for stem wood volume measurement are an important part of stem biomass determination.

This chapter will consider the various ways in which the wood volume of individual tree stems or logs is measured.

5.2   'Exact' Volume Measurement

No tree stem is perfectly regular in shape. All stems have bends, twists and lumps where branches have emerged or there have been other environmental influences that have affected stem shape (Sect. 3.2). Despite this, there are at least four methods of measurement that can take account of this complexity of stem shape and, thus, can provide very precise estimates of stem volume with virtually no bias.

The first method involves immersion of the stem (perhaps after cutting it into sections) in water and measurement of the volume of water displaced. This is known as xylometry. Generally, it is impractical for any but exacting research work. It requires large immersion tanks that are not portable for field use and the tree must be felled before it can be measured. There are various examples of the use of xylometry in research projects (Martin 1984; Filho and Schaaf 1999; Figueiredo et al. 2000; Özçelik et al. 2008b).

The second method uses lasers (Sect. 2.2) and has become common in sawmills to assist in determining the optimal set of timber products that can be sawn from a

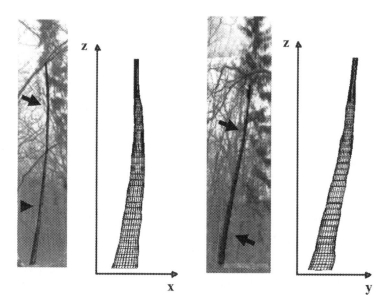

Fig. 5.1 Stem profile of a standing 70-year-old, Sitka spruce (*Picea sitchensis*) tree, derived from digital photographs of the tree. The total height of the tree was 22 m and its diameter at breast height over bark was 29 cm. Photographs were taken from two directions at right angles and used to produce the line drawings shown from which a three-dimensional profile of the stem was obtained (from Fig. 4 of Hapca et al. 2007 and reproduced by kind permission of the Annals of Forest Science)

log. As a log enters the sawing line, multiple lasers scan it from several directions and measure its diameter at very short intervals along its whole length. This information is processed by a computer to produce a precise, three-dimensional profile of the log from which its volume can be determined. Because of the complex laser equipment involved, this method is not practical for field use and, as with xylometry, requires that the tree be felled before measurement.

The third method can be used with standing trees in the field. Using a digital camera, a tree is photographed from at least two directions. With computer analysis, a three-dimensional view of its stem may then be produced; an example from Hapca et al. (2007) is shown in Fig. 5.1. Using a computer, the volume of the tree stem could then be determined, taking account of the fine detail of any irregularities along it. However, a standing tree still has bark on its stem, so the volume could only be inclusive of the bark and wood together. In the example in Fig. 5.1, the tree did not have a very dense crown, so the camera was able to picture the stem well up into it; this would not be the case for many other types of trees that have much denser crowns. Other research has been developing the use of photography to achieve equivalent results (Dean 2003; Dean and Roxburgh 2006; Morgenroth and Gomez 2014).

A fourth method, analogous to digital photography, involves laser scanning of a standing tree from an instrument placed on the ground at some distance from the tree. Such instruments are described in more detail in Sect. 13.1. Both laser scanning and photography are methods known by the general term **remote sensing**, that is, measurement methods that rely on equipment that takes measurements of objects at some distance from them. The use of remote sensing is a rapidly developing area of tree and forest measurement and is discussed in greater detail in Chap. 13.

5.3 Volume by Sectional Measurement

It is common in forestry practice to need to measure tree stem wood volumes in the field. Whilst new methods of remote sensing are being developed to aid in this (Sect. 5.2, Chap. 13), such measurements are still often made directly by people. The methods they use have a long history of development going back to the nineteenth century. They are not able to take as much account of the irregularities in shape along a tree stem as the methods described in Sect. 5.2. Fortunately, however, most trees in most forest circumstances are sufficiently regular in shape that these older methods can measure stem volumes with an accuracy and level of precision that is adequate for most forestry purposes. The methods can be destructive (the tree is felled before measurement) or non-destructive (the tree is measured standing).

The principal one of these methods is known as the sectional method. It involves measuring a tree stem in short sections, determining the volume of each section and summing them to give the total volume.

5.3.1 Sectional Volume Formulae

In the sectional method, the volume, V_S, of a section of a stem is determined by measuring the length of the section, l, and some or all of the stem diameter at the lower end of the section (commonly referred to as the large end diameter), d_L; the diameter at the upper end of the section (small end diameter), d_U; and the diameter midway along the section, d_M. These measurements are used to determine the volume of the section using one of three formulae, each named after the person who first developed it. They are Smalian's formula,

$$V_S = \pi l \left(d_L{}^2 + d_U{}^2 \right)/8,\qquad(5.1)$$

Huber's formula,

$$V_S = \pi l d_M^2 / 4, \tag{5.2}$$

and Newton's formula,

$$V_S = \pi l \left(d_L^2 + 4 d_M^2 + d_U^2 \right) / 24. \tag{5.3}$$

The units of the measurements used with these formulae must be consistent, say, all in metres or all in feet. So, for a 3 m long stem section with $d_L = 0.320$ m, $d_M = 0.306$ m and $d_U = 0.296$ m, its volume estimated by Smalian's formula (5.1) would be 0.224 m^3, by Huber's formula (5.2) 0.221 m^3 and by Newton's formula (5.3) 0.222 m^3. The differences in the results arise from the different amounts of information used to calculate each and natural irregularities along the stem section.

These three formulae have been an integral part of forest measurement for many years and remain so today. All three will give an unbiased estimate of the volume of a stem section if the section is cylindrical or shaped as part of what is known as a quadratic paraboloid (Sect. 5.3.2). Newton's formula will give an unbiased result also if the stem section is shaped as part of a cone or a neiloid (see Sect. 5.3.2). Of course, even if a stem section is shaped generally like one of these specific shapes, irregularities along the stem (Sect. 3.4) will ensure that none of these formulae can be expected to give a section volume exactly.

In practice, Newton's formula has generally the least potential for bias and has the highest precision because it involves three measurements of each stem section. Huber's formula is then generally more reliable than Smalian's because sections that have flared ends (such as those in the butt swell region) can cause deviant results in the latter case. However, Smalian's formula is perhaps used most commonly because mid-diameter, as required for Huber's and Newton's formulae, is difficult to reach when logs are piled and stems are not always measured at regular intervals so that mid-diameters are included as part of the measurement sequence (Fonweban 1997; Figueiredo et al. 2000; Avery and Burkhart 2002, p. 102).

Whilst other formulae and indeed different methods have been developed from time to time to be used as alternatives to (5.1)–(5.3) (Wiant et al. 1991; van Laar and Akça 2007; Özçelik et al. 2008b), none is in use consistently today and will not be considered further here.

5.3.2 Tree Stem Shape

As discussed above, Smalian's, Huber's and Newton's formulae assume that tree stems have particular shapes. To understand how these formulae have become such an important part of forest measurement practice, it is necessary to consider how tree stems are shaped. Only then will it be possible to judge how appropriate these formulae really are.

Tree stem shape can be defined as the way in which stem diameter changes with height along the stem. Much research was undertaken in the pre-computer era of the twentieth century to try to determine how tree stems are shaped. Summarising that research in modern parlance, it was believed that the stem diameter, d_x, at any distance x from the tip of a stem could be described by the relationship

$$d_x = \kappa x^\rho \tag{5.4}$$

where κ and ρ are **parameter**s of the equation, that is, variables that take particular values in the equation for a particular stem from which has been measured a set of stem diameters and distances from the tip. Note that Greek letters have been used to represent the parameters of this equation. Their names are listed in the Greek alphabet given in Appendix 3.

The older research suggested that tree stem shape varied in different parts of the stem. It was believed that near the base of the tree stem, in the region where the butt swell occurred, the stem generally had a shape known as a neiloid, when the parameter ρ in (5.4) has the value 1½. Above that, and for the main part of the tree stem, at least into the lower part of the crown, it was believed that the stem was shaped as a quadratic paraboloid, when $\rho = \frac{1}{2}$ in (5.4). The top section of the stem was believed to be conical, when $\rho = 1$ in (5.4). Since the main part of the stem was believed to be shaped as a quadratic paraboloid, and particularly because that is the part of the stem that is most used for timber, this led to the use of (5.1)–(5.3), all of which will give an unbiased estimate of volume if the stem is indeed shaped as a quadratic paraboloid.

The advent of the computer has allowed much more detailed analysis of tree stem shape. In particular, it has been found that tree stems vary their shape more or less continuously along their length. Functions much more complex than (5.4), known as **taper function**s (Chap. 6), have been developed to describe stem shape.

This is illustrated in Fig. 5.2. There, the shape of the stem under bark is shown for a typical blackbutt (*Eucalyptus pilularis*) tree, a species important for wood

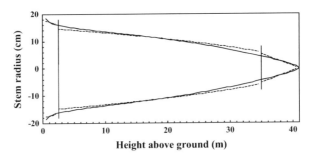

Fig. 5.2 Stem profile under bark of a felled blackbutt (*Eucalyptus pilularis*) stem (*continuous line*), with a diameter at breast height over bark of 40 cm and a total height of 41 m. The *vertical solid lines* are positioned at heights of 2.5 and 35 m above ground. The *dashed lines* show what the stem profile would be if the stem was shaped as a neiloid between 0 and 2.5 m height, a quadratic paraboloid between 2.5 and 35 m and a cone between 35 and 41 m. Note that the horizontal and vertical scales of the diagram differ greatly (drawn using an unpublished taper function, developed in 1997 by E. Baalman, then of the state forest service of New South Wales, Australia)

production in subtropical Australia. That shape was drawn using a taper function developed for that species. Of course, taper functions only show a smoothed stem, without the minor irregularities that will occur naturally in any real stem (Sect. 3.4).

Superimposed as dotted lines on the tree stem shape shown in the figure are the shapes that the stem would have *if* its lowest 2.5 m was shaped as a neiloid, *if* the main part of the stem between 2.5 and 35 m was shaped as a quadratic paraboloid and *if* the last 6 m of the stem was conical in shape. Whilst some deviations are apparent between these three specific shapes and the actual stem shape, the differences are not great. That is, the findings of the older research do not seem to have been violated too grossly. This has justified the continued use of (5.1)–(5.3) to determine the volumes of sections of tree stems or logs.

5.3.3 Sectional Measurement of Felled Trees

It is quite straightforward to use the sectional method to measure felled trees or individual logs lying on the ground. In both cases, the bark can be removed so that the stem wood is measured directly. Removal of bark can be quite difficult, especially at times of year when the trees are not growing rapidly and the bark may be held very tightly on the stem. However, once the bark is removed, stem diameter is measured with a diameter tape or calipers at successive distances along the stem right to the tip of the stem or to whatever stem diameter at which it is desired to stop measuring.

Various decisions need to be made when taking these measurements. If Huber's formula (5.2) is to be used to calculate the volume of each section, only section mid-diameters are measured. If Smalian's (5.1), the diameters at both ends are measured. If Newton's (5.3), all three diameters must be measured. Because of the butt swell, care needs to be taken near the base of the tree. The enlarged diameter at the base of the stem can lead to substantial bias in the estimate of the volume of the base section if that diameter is used with Smalian's or Newton's formulae. Care is needed also if the end of a section falls on a bump on the stem, such as where a branch has emerged. The usual technique of averaging two diameter measurements spaced equally above and below the irregularity must then be used (Sect. 3.4).

Perhaps more importantly, a decision needs to be made as to what length the sections should be. Mathematically speaking, the shorter and shorter the sections, the less and less important become any violations of the assumptions about stem shape that are implicit in (5.1)–(5.3) and the less will be any bias in the volume estimates made using them. Most direct tree stem volume measurements today are made using quite short section lengths, perhaps of 0.5–1.5 m for large trees. Shorter lengths still are necessary for small trees only a few metres tall.

Less attention is usually paid to finding the volume of the last several metres of the stem of a large tree, perhaps above the point where the stem diameter falls below 3–5 cm, than for the lower part of the stem. The uppermost part of the stem contributes very little to the total volume of the stem, so a less precise measurement of it is unimportant in adequately determining the entire stem volume. Usually, only the diameter of the stem at the base of that last section, d_T, is measured together with the distance, t, to the tip of the stem. Its volume, V_T, is then determined as that of a cone, that is,

$$V_T = \pi t d_T^2 / 12 \qquad (5.5)$$

where d_T and t must have the same units.

Sometimes, it is desired to determine the volumes of particular classes of log size that occur in a tree. Nothing new in the approach is necessary to do this. Still, care is needed to make sure that the stem is divided into sections that define the different log size classes so that their respective volumes can be determined. Section lengths may then become somewhat irregular. Smalian's formula is usually the easiest to apply to obtain the section volumes, because the position of the midpoint of each section does not have to be located.

In modern forestry practice, it has become quite rare to need to measure the stem wood volumes of individual trees using the sectional method as part of routine **forest management**. Mostly, it is done as part of a research project to obtain the data necessary to develop a tree **volume function** or taper function for use with a particular tree species in some region of forestry interest. These functions allow estimation of stem wood volumes of standing (or felled) trees, usually from measurement only of their diameter at breast height over bark and their total height. They will be discussed in detail in Chap. 6.

5.3.4   Sectional Measurement of Standing Trees

Standing trees could be climbed and measured sectionally (measuring both diameter over bark and bark thickness) to determine their stem wood volumes. However, this is dangerous and labour-intensive work that is done rarely today. As mentioned above, tree volume and taper functions (Chap. 6) are available now for many forest tree species in many parts of the world to provide (usually quite precise) estimates of their wood volume from measurements of them that can be made easily from the ground.

Generally, it is only in circumstances where adequate volume or taper functions are not available that stem wood volume of standing trees must be measured directly. Where it is necessary, various high-quality optical instruments are available (e.g. the Barr and Stroud **dendrometer**, the Telerelaskop, the Relaskop, the Wheeler pentaprism, the laser relascope—see Kalliovirta et al. 2005) that allow stem diameters over bark to be measured from the ground at any height desired along the stem. Using such instruments to make these measurements is known as optical dendrometry. Accuracies of measurement with these devices may be to within 1–5% of the actual stem diameter.

Because optical dendrometers measure diameters over bark only, some additional method must be used to determine bark thicknesses, so that wood volume under bark can be determined. An example of the type of method used to do this is given in Sect. 5.4. Muhairwe (2000) has summarised and tested a variety of functions that have been developed from time to time for this purpose.

One difficulty with these instruments is to sight the stem clearly amongst the branches in the crown. Section lengths need to be adjusted to make measurements at heights where the stem can be seen. Care and practice are needed to use these instruments efficiently; they are slow and time consuming but much less arduous and dangerous than climbing trees. Another measurement method that still uses an optical dendrometer, but requires much less intense measurement, is discussed in Sect. 5.4.

5.4   Volume by Centroid Sampling

To measure stem volume of standing trees, a method known as centroid sampling involves much less measurement than the use of sectional measurement with optical dendrometry (Sect. 5.3.4). Firstly, it requires that the stem diameter at breast height, both under and over bark, and total height of a tree be measured. Then, one further value of stem diameter over bark must be measured, usually high up on the tree stem. Whilst research has identified other methods that require similar measurements to obtain stem volumes of standing trees and studied their application (Forslund 1982; Wiant et al. 1991, 1996; Van Deusen 1987, 1994; Van Deusen and Lynch 1987; Yamamoto 1994; Özçelik 2008; Özçelik et al. 2008a, 2010; Ducey and Williams 2011), none of those has attracted as much consideration as centroid sampling and they will not be considered further here.

Not only can centroid sampling be used to measure the total stem wood volume of a standing tree, but it can be used also to measure the wood volume of any section of the tree stem between any two heights above ground that are desired. This would allow measurement of the volume of a section of the stem from which a log of any particular size might be cut. However, an additional upper stem diameter must be measured to determine the volume of each particular stem section of which the volume is required.

Centroid sampling proceeds as follows (Furnival et al. 1986; Gregoire et al. 1986; Wiant et al. 1992b; Gregoire and Valentine 2008, pp. 111–112). Suppose a tree is measured as having a diameter at breast height over bark of D_o and its corresponding diameter under bark at breast height is measured as D_u. Suppose its total height (ground to tip) is measured as H. Suppose it is desired to estimate its stem wood volume under bark between two points on the stem, from a lower height, H_l, to an upper height, H_u, that is, $0 \leq H_l < H_u \leq H$.

The stem diameter over bark must now be measured at a point on the stem somewhere between H_l and H_u. The height at which this measurement is to be taken, H_s, is

$$H_s = H - \left[(H - H_1)^2 - \eta k\right]^{\frac{1}{2}},$$ (5.6)

where

$$k = 2H(H_u - H_1) + H_1^2 - H_u^2.$$ (5.7)

When this method was being developed, the value used for η in (5.6) was a randomly selected value in the range 0–1; a different value was used for each tree being measured. Under these circumstances, the method is known as importance sampling. In centroid sampling, the value of η is fixed at ½. This determines the position of the centroid of the stem section being considered; one half of the volume of the section lies above and below the centroid position. The computations in (5.6) and (5.7) would be done with a calculator or computer at the time a tree was being measured in the field.

Once H_s has been calculated, the diameter over bark of the stem must be measured at that height; when this is high up on the stem, the measurement would be made from the ground using an optical dendrometer (Sect. 5.3.4). Suppose this diameter is measured as D_s. Then, the wood volume under bark of the section of the stem between H_1 and H_u, V_{lu}, may be estimated as

$$V_{lu} = \pi k (D_s D_u / D_o)^2 / [8(H - H_s)]$$ (5.8)

Note that the same units must be used throughout for each of these measured variables (say, all in metres or all in feet).

Two important assumptions have been made in deriving (5.6)–(5.8). The first is that the ratio of diameter under bark to diameter over bark is constant anywhere along the stem. Research with many forest tree species has suggested this is often so, at least for a large part of the stem. This assumption could be used to determine under bark diameters from over bark when the sectional method is used with optical dendrometry (Sect. 5.3.4); other methods are available to do this (Muhairwe 2000; Leites et al. 2013) and their assumptions could replace this assumption in the present theory.

The second assumption is that the stem is shaped as a quadratic paraboloid along its whole length. As discussed in Sect. 5.3.2, this is true generally for a large proportion of the tree stem, above the butt swell and below the crown. The method can be modified for cases where research has provided better information about tree stem shape (Wiant et al. 1989; Wood et al. 1990; Coble and Wiant 2000).

Consider an example of the use of centroid sampling. Suppose a tree had $D_o = 0.423$ m, $D_u = 0.373$ m and $H = 38.0$ m. Suppose an estimate of stem wood volume was required for the section of the stem from $H_1 = 0.2$ m above ground to $H_u = 8$ m above ground. Using (5.7), this gives a value of $k = 528.84$. With $\eta = ½$ in (5.6), then the height at which the upper stem diameter over bark must be measured would be $H_s = 3.9$ m. Suppose that diameter was then measured as $D_s = 0.333$ m.

Then, using (5.8), the wood volume of the required stem section would be estimated as 0.525 m^3.

The method can be used to estimate total stem wood volume from ground to tip as well as the volume of any section of the stem. Considering the same example tree as above, this would involve setting $H_l = 0$ m and $H_u = H = 38.0$ m. This would lead to $H_s = 11.1$ m and the stem diameter over bark would then be measured at that height. Suppose this diameter was 0.243 m, then the total stem wood volume would be estimated as 0.969 m^3.

A number of research studies have considered in detail how well both importance and centroid sampling perform in practice. These have compared estimates of stem volumes of a wide range of both **softwood** and **hardwood** tree species to volumes determined from detailed sectional measurements (Sect. 5.3) or by xylometry (Sect. 5.2). Wiant et al. (1989, 1996) found there was negligible bias when using importance sampling. Other works found that centroid sampling sometimes showed a tendency to underestimate stem volumes, although generally by appreciably less than 5 % and often with little statistical certainty that the underestimation was real (Wood and Wiant 1990, 1992a, b; Wood et al. 1990; Wiant et al. 1991, 1992a, 1996, 2002; Coble and Wiant 2000; Özçelik 2008; Özçelik et al. 2008a, 2010). However, centroid sampling was generally a more precise estimator of stem volume than importance sampling (Wood et al. 1990; Wood and Wiant 1992a; Wiant et al. 1996). Another advantage of centroid sampling was that the point on the stem at which the upper diameter measurement needed to be made was rarely within the tree crown, whereas, with importance sampling, it may be partly hidden within the crown. Further work has explored how, under some circumstances, it might be possible to use centroid sampling and avoid problems that might arise because of the slight bias involved with it (Williams and Wiant 1998).

These research studies have concluded generally that centroid sampling is an adequate method for estimating the wood volume of part or all of the stems of standing trees and is generally to be preferred to importance sampling. However, its use in practice seems to have been limited; two examples can be found in Dieters and Brawner (2007) and West et al. (2008).

Use of centroid sampling, rather than using a volume function or taper function (Sect. 6.3), will probably improve precision of the volume estimate that is being made; Özçelik et al. (2008a, 2010) gave an example where this was so. Wiant et al. (2002) found that using pre-existing taper functions for a number of North American hardwood species yielded appreciably more biased estimates of standing tree stem volume than using centroid sampling. Of course, further research work will be necessary to confirm that centroid sampling is both sufficiently unbiased and improves precision sufficiently that its use is worthwhile for any particular forest type being considered.

Chapter 6
Stem Volume and Taper Functions

Abstract Direct measurement of the wood volume in a tree stem is a difficult and time-consuming process, particularly if the tree is standing. For many forest species around the world, functions have been developed that allow estimation of the volume in all of the stem or that much of it which is large enough to be merchantable. These 'volume' functions require measurement only of the diameter at breast height over bark and total height of the tree and, sometimes also, the degree of taper of the stem determined by diameter measurement at two heights. More recently, functions have been developed widely across the world that allow estimation of the stem diameter at any height up a tree stem, again, usually from measurement only of diameter at breast height over bark and total height of the tree. Whilst rather more complex mathematically than volume functions, these 'taper' functions allow estimation both of the total wood volume in a tree stem and of the dimensions of the individual logs that might be cut from the stem as required by the market that the forest is supplying. This chapter describes these functions and their application.

6.1 The Functions

As discussed in Chap. 5, stem volume measurement is a laborious and time-consuming task even for felled trees. In modern forestry practice, one of the most common reasons for taking such measurements is to develop stem volume functions or taper functions for a particular tree species in a particular forest region.

Volume functions allow estimation of the total stem volume of a standing tree from simple measurements, usually its diameter at breast height over bark and its total height. Sometimes also they allow estimation of the volume of parts of the stem that are large enough to be sold as logs, that is, the **merchantable volume** of wood available from the tree.

Taper functions estimate how the diameter of a tree stem varies along its length. As with volume functions, they generally require that only the diameter at breast height and total height of the tree be measured. Furthermore, they can be used to estimate both merchantable and total stem wood volumes. In particular, they can estimate in much greater detail than volume functions the dimensions of individual

© Springer International Publishing Switzerland 2015

P.W. West, *Tree and Forest Measurement*, 3rd edition,

DOI 10.1007/978-3-319-14708-6_6

logs that can be cut from a stem as required by the market that the forest is supplying. Whilst taper functions are rather more complex to develop than volume functions, the data required for both is the same: sectional measurements of the stems of a large **sample** of individuals of the tree species concerned usually provide the data required.

Volume and taper functions have been developed for a very large number of the tree species important commercially to forestry around the world. In this chapter, these functions will be described generally, with reference to just a few examples.

6.2 Volume Functions

Three variations of stem volume functions will be considered. Firstly, there are those that allow estimation of total stem volume from ground to tip from measurements of diameter at breast height and tree total height. Secondly, there are variations on those functions that require that a measure of the degree to which the stem tapers be made also. Thirdly, there are functions that allow estimation of the volume of part of the stem so that merchantable volume can be estimated.

6.2.1 Volume from Diameter and Height

Different researchers, working with different species in different parts of the world, have used various **functional form**s for stem volume functions. The most common form can be described by the general equation

$$V = \alpha + \beta D^{\gamma} + \delta D^{\varepsilon} H^{\zeta} + \cdots, \tag{6.1}$$

where V is total stem volume from ground to tip over or under bark; D is stem diameter at breast height, usually over bark, and H is tree total height. The terms α, γ, δ, ε and ζ are parameters of the equation that will take particular values in the function developed for a particular species in a particular region. The string of dots after (6.1) implies that additional terms have sometimes been included in functions of this form, usually terms in specific powers of D and H and their products. Research continues on which functional forms are most appropriate for stem volume functions in particular cases (Williams and Schreuder 2000; Hjelm and Johansson 2012).

Given that a data set has been collected from a sample of trees of a species, for each of which V, D and H have been measured (volume usually by direct measurement using the sectional method), the values of the parameters of an appropriate volume function will usually be determined using regression analysis. Regression analysis is one of the most powerful tools available for the analysis of data in the natural and physical sciences. It is used to determine how variables are related to

Table 6.1 An arbitrary selection from the literature of stem volume functions that allow total stem volume from ground to tip under (V_U, m^3) or over bark (V_O, m^3) to be estimated for an individual tree from its diameter at breast height over bark (D, m) and total height (H, m)

Number	Function	Species and location	Reference
1	$V_U = 0.298D^2H$	*Eucalyptus regnans*, New Zealand	Hayward (1987)
2	$V_U = 0.001 + 0.270D^2H$	*Eucalyptus viminalis*, New South Wales, Australia	Bi (1994)
3	$V_U = 0.005 + 0.330D^2H$	*Pinus taeda*, Southern USA	Williams and Gregoire (1993)
4	$V_U = 0.037 + 0.28D^2H$	*Bursera simaruba*, Puerto Rico	Brandeis et al. (2006)
5	$V_U = 0.250D^{1.85}H^{1.03}$	*Picea glauca*, Canada	Morton et al. (1990)
6	$V_U = 0.128D^{1.69}H^{1.16}$	*Pseudotsuga menziesii*, USA and Canada	Rustagi and Loveless (1990)
7	$V_O = 0.450D^{1.92}H^{0.90}$	*Juniperus procera*, Ethiopia	Pohjonen (1991)
8	$V_O = 0.311D^{1.93}H^{1.02}$	*Gironniera subaequalis*, Southern China	Fang and Bailey (1999)
9	$V_O = -0.041 + 5.3D^{1.90} + 9.23 \times 10^{-7}D^{0.775}H^{4.18}$	*Populus* spp., Sweden	Hjelm and Johansson (2012)
10[a]	$V_U = \exp[-1.75 + 1.29/(D+1.27)^2]D^2H$	*Eucalyptus regnans*, Victoria, Australia	Opie (1976)

[a]The expression 'exp' in this function means that the mathematical constant 'e' is to be raised to the power of the value calculated in the square brackets []. The constant 'e' is the base of natural logarithms. Its value is approximately 2.7183

each other and to provide predictions of values of one variable from one or several other variables. There will be no further discussion here of how regression analysis works, other than to say that some knowledge of mathematical statistics and some years of experience are necessary to apply it competently. It is discussed in standard texts (e.g. Draper and Smith 1988; Freund et al. 2006). The calculations required to do it are rather tedious and are done using a computer.

Table 6.1 lists an arbitrary selection from the literature of ten stem volume functions developed for various tree species in various parts of the world. All but the last of these functions have a form derived from (6.1). The tenth function has a quite different form and was included to illustrate that individual authors have used quite different functional forms from time to time. The specific values that the parameters took for each function are shown in the table.

Examination of these equations will show that different tree species may have substantially different stem volumes for any given stem diameter and height. This occurs because the shapes of the stems of different species can differ appreciably. It means that applying a stem volume function developed for one species to another species could lead to substantial bias in the volume estimates obtained for the other species. For this reason, forest scientists have developed many volume functions for many tree species and often for particular regions where a species occurs. Work continues on the development of such functions (Cordero and Kanninen 2003; Akindele and LeMay 2006; Brandeis et al. 2006; Diéguez-Aranda et al. 2006b; Vallet et al. 2006; Khan and Faruque 2010; Henry et al. 2011; Fonweban et al. 2012; Hjelm and Johansson 2012).

As a 'rule of thumb', the volume function

$$V_U = 0.3D^2H, \tag{6.2}$$

where V_U is tree stem volume under bark from ground to tip (m^3), D is tree diameter at breast height over bark (m) and H is tree total height (m), may be used to give a rough estimate of tree stem wood volume for any tree of given diameter and height. Of course, I would not advocate use of this function if it was desired to estimate properly the stem volume of any tree. However, it is easy to remember and allows a rough estimate to be made of the stem volume of a tree when a specific volume function for that species is not to hand.

The volume functions described so far have required that only tree stem diameter at breast height over bark and tree total height be measured. However, other functions have been developed that require measurement also of the taper of the stems, that is, how its diameter changes with distance along the stem.

The degree to which tree stems taper differs from species to species. It varies also between individuals of any one species, depending on their circumstances. Wind exposure is a major influence. The forces to which tree stems are subjected as they sway in the wind cause their stems to become more tapered. In an experiment where large trees were tied with cables to prevent them swaying, the tree stems became almost cylindrical after several years, that is, to have virtually no taper (Jacobs 1954). Other experiments have shown that bending stresses in the stem due to wind sway increase both wood strength and stem taper (Valinger 1992; Osler et al. 1996a; Dean et al. 2002; Mäkelä 2002; Fourcaud and Lac 2003; Fourcaud et al. 2003; Dean 2004; Watt et al. 2006a, b). At present, our understanding of this process is inadequate and no theory has yet been developed to describe fully how and why stems taper.

In forestry, management practices are often undertaken that affect the degree to which trees are exposed to the wind and, hence, the degree to which they sway. Opening the forest, by removing some of the trees in thinning a plantation, is one of the most common of these practices. Under those circumstances, it would be expected that a stem volume function that took account of stem taper might be better able to estimate stem volumes in both thinned and unthinned stands.

The additional information required to assess stem taper usually involves measuring another tree stem diameter at some height other than breast height. The height chosen is usually not far up the stem, so that the additional diameter can be reached without too much difficulty. Two examples will be used to illustrate such functions. The first was developed by Rustagi and Loveless (1990) for Douglas fir (*Pseudotsuga menziesii*) in North America. The function was

$$V_{\mathrm{U}} = 0.146 D^2 H + 0.433 D^2 H_{\mathrm{D}}, \qquad (6.3)$$

where V_{U} (m^3), D (m) and H (m) were as for (6.2) and H_{D} (m) was the height up the tree stem at which the diameter over bark was two thirds of the diameter at breast height over bark (a height that may be rather tedious to locate on a tree). Stem taper was being allowed for through the variable H_{D} in this function; the more the stem tapered, the less would be that height. Rustagi and Loveless found that this function gave appreciably more precise estimates of tree stem volume than a function that did not include the measure of stem taper (it was Function 6 in Table 6.1).

The second example was developed by Aleixo da Silva et al. (1994), using combined data for loblolly pine (*Pinus taeda*) and slash pine (*Pinus elliottii*) from the southern USA and Honduran Caribbean pine (*Pinus caribaea* var. *hondurensis*) from Sri Lanka. Their function was

$$V_{\mathrm{O}} = 0.196 D^2 H [1 + D_{\mathrm{a}}/D_{\mathrm{b}}][H/(H - 1.52)] \qquad (6.4)$$

where V_{O} was tree stem volume over bark from ground to tip (m^3), D (m) and H (m) were as above and D_{a} and D_{b} were the stem diameters over bark (m) at 1.52 and 0.152 m above ground, respectively. In this case, it was the ratio of the two diameters from two different heights on the stem, D_{a} and D_{b}, that acted as the measure of stem taper. It is interesting that this single function seemed to give reliable results for three different species of pine in two very distant parts of the world.

6.2.2 Merchantable Stem Volume

The volume functions described in Sect. 6.2.1 provide estimates only of the total volume of the stem. This is useful to forest scientists as a measure of tree growth and, as will be seen in Sect. 7.4.2, can be useful also in estimating tree biomasses. However, in forestry practice it is often desired to estimate the volumes available of logs of various size classes that can be sold from a tree, that is of the merchantable volume available.

Given the diameter at breast height over bark and the total height of a tree, merchantable volume functions usually estimate the volume up to some particular diameter along the stem that determines the point above which logs of a particular size class can no longer be cut. An example comes from Shiver and Brister (1992) for plantations of Sydney blue gum (*Eucalyptus saligna*) up to about 10 years of age in Kenya. Their function was

$$V_{\mathrm{d}} = 0.0950 D^{1.83} H^{1.24} \left[1 - \left(1.24 d^{3.49}/D^{3.37}\right)\right], \qquad (6.5)$$

where V_d was the stem volume under bark (m³) from ground level up to the point on the stem where its diameter over bark became d (m), D was tree diameter at breast height over bark (m) and H was tree total height (m).

Consider how this function might be useful. Suppose that two possible products could be cut from trees in these forests, posts or **pulplogs** (small logs to be chipped and used for papermaking). Suppose posts must have a diameter over bark at their small end of at least 0.15 m and pulplogs can be cut up to a stem diameter over bark of 0.06 m. Consider a tree with $D = 0.24$ m and $H = 25$ m. Then, using $d = 0.15$ m, (6.5) shows that 0.301 m³ of the stem could be cut as posts. With $d = 0.06$ m, the function shows 0.375 m³ of pulplogs could be cut from the stem. If all the volume that could be used for posts was indeed used, then there would be 0.074 m³ ($=0.375 - 0.301$) left to be sold as pulplogs. This type of information would be useful to forest planners attempting to determine what volumes of posts and pulplogs could be harvested from trees in this forest.

One of the limitations of this type of merchantable volume function should be evident from this example. Merchantable log sizes are defined usually not only by the minimum diameter they may have at the small end of the log but also by the minimum length the log must have. Thus, it might be that a post must be at least 2 m long, say, or it would be too short to be sold as a post. Whilst the tree in the example would yield a total of 0.301 m³ of material large enough in diameter to be posts, there might be insufficient length of material in that volume for it all to be used as posts. No information about the length available is given by the merchantable volume function. As discussed in Sect. 6.3, taper functions overcome this limitation.

Merchantable volume functions have been developed also that predict volume up to some diameter, like (6.5), or volume up to some height or the ratio to total stem volume of volume up to some diameter or height. These functions can then be combined to form a *de facto* taper function to overcome this limitation of merchantable volume functions. Teshome (2005) and Fonweban et al. (2012) gave examples of this.

It should be noted also that if $d = 0$ in the example above, that is, where the stem diameter becomes zero at its tip, (6.5) then gives the total stem wood volume of the tree from ground to tip. In other words, (6.5) can be used in just the same way as volume functions that estimate total stem volume only, functions such as those in Table 6.1.

An additional problem in determining merchantable volume (or indeed total stem wood volume) is that trees are found quite often with their stem partially hollowed out. Termites and wood decay fungi are the most common causes of this. The damage may extend for several metres up from the base of the tree reducing substantially the amount of usable wood that can be cut from it. Externally, the tree may appear perfectly sound.

If the presence of such damage is suspected, it can be assessed using instruments that can examine the stem interior remotely (Martinis et al. 2004; Butnor et al. 2009; Brazee et al. 2011). However, its presence often does not become evident until the tree is felled. Functions have been developed that can at least estimate the likelihood that a stem of a tree of particular species contains such damage; Schneider et al. (2008) gave a good example.

More subtle degradation of wood quality can occur where wood is stained through fungal or microbial action; this leads only to degradation of its appearance, not to loss of wood or reduction in its strength. Baral et al. (2013) gave an example of this in sugar maple (*Acer saccharum*), an important timber tree in North America. They developed a model system to estimate the amount of discoloured wood standing tree stems contained based on tree age, external size, number of injuries apparent on the stem and the productivity of the site on which they were growing.

6.3 Taper Functions

Taper functions estimate how the diameter of the stem (over or under bark) changes along the length of the stem. Because understanding of how and why stems taper is inadequate at present (Sect. 6.2.1), most taper functions developed to date are **empirical** functions. That is, they have been determined by trial and error by first measuring how diameter changes along the length of tree stems and then finding some mathematical function that describes adequately the stem shape. Research has been attempting to develop taper functions from a more theoretical basis, but this work still has far to go (Sharma and Oderwald 2001; Deleuze and Houllier 2002; Ikonen et al. 2006).

In this section, some examples will be given of taper functions that have been developed for various species in various parts of the world. The way in which they can be used to estimate both total and merchantable stem volumes will then be described.

6.3.1 Examples of Taper Functions

Sharma and Oderwald (2001) developed a taper function for trees in natural forests of loblolly pine (*Pinus taeda*) in the southern USA. It was

$$d_{oh} = D\left[(h/1.37)^{-0.185}(H - h)/(H - 1.37)\right]^{\frac{1}{2}}, \tag{6.6}$$

where d_{oh} was stem diameter over bark (cm) at height h (m) above ground, D was stem diameter at breast height over bark (cm) and H was tree total height (m). Malimbwi and Philip (1989) developed a function for plantations of Mexican weeping pine (*Pinus patula*) in Tanzania as

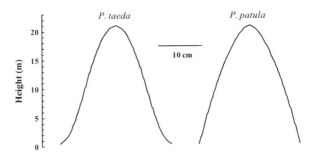

Fig. 6.1 Profiles of the stems of trees of *Pinus taeda* over bark (as predicted by 6.6) and of *Pinus patula* under bark (6.7). In both cases, the trees were assumed to have a diameter at breast height over bark of 23 cm and a height of 21 m. Note that the vertical and horizontal scales of this figure differ greatly (drawn using taper equations of Sharma and Oderwald 2001 and Malimbwi and Philip 1989)

$$d_{uh} = 0.774 D^{0.932} (H - h)^{0.610} / H^{0.448}, \qquad (6.7)$$

where d_{uh} was stem diameter under bark (cm) at height h (m) above ground.

Figure 6.1 shows the stem shapes predicted by these functions for example trees of a particular diameter and height. The stem profile for *Pinus taeda* shows clearly the butt swell near the base of the stem, but the butt swell is missing for *Pinus patula*; Malimbwi and Philip did not describe fully how they collected the data for their *Pinus patula* trees, and it may be that they excluded lower stem data.

Mathematically speaking, both (6.6) and (6.7) are quite simple. Functions such as these often describe tree stem shape quite well, particularly in its midsections. However, rather more complex functions seem to be necessary to ensure the shapes of the upper and lower sections of the stem are defined well also. As an example, Brooks et al. (2008) used a function suggested by Max and Burkhart (1976) to describe the stem shape of three commercially important species native to Turkey, Brutian pine (*Pinus brutia*), Cedar of Lebanon (*Cedrus libani*) and Cilicica fir (*Abies cilicica*). Sharma and Burkhart (2003) used the same function for loblolly pine (*Pinus taeda*) from plantations in the south-eastern USA. This rather cunning function considers the tree stem to be shaped as a neiloid near its base, as a paraboloid in its midsection and as a cone near the tip, that is, consistent with the older research findings discussed in Sect. 5.3.2. The function is

$$d_{oh} = D \left\{ b_1 [(h/H) - 1] + b_2 [(h/H)]^2 - 1 + b_3 [a_1 - (h/H)]^2 I_1 + b_4 [a_2 - (h/H)]^2 I_2 \right\}^{\frac{1}{2}},$$
$$(6.8)$$

where d_{oh} (cm), h (m), D (cm) and H (m) have the same meanings as above, a_1, a_2, b_1, b_2, b_3 and b_4 are parameters, $I_1 = 1$ when $(h/H) \leq a_1$ or zero otherwise and $I_2 = 1$ when $(h/H) \leq a_2$ or zero otherwise; note that in Sharma and Burkhart's case, they predicted diameter under bark (d_{uh}, cm) rather than diameter over bark. In (6.8), the

Table 6.2 Parameter values of (6.8) for four tree species

Parameter	Pinus brutia	Cedrus libani	Abies cilicica	Pinus taeda
a_1	0.7313	0.7593	0.849	0.7487
a_2	0.1307	0.1116	0.171	0.0867
b_1	−3.0832	−3.6549	−2.9364	−3.3108
b_2	1.486	1.7947	1.3965	1.5745
b_3	−0.9304	−1.3658	−0.7093	−1.7299
b_4	17.9703	25.9476	5.4083	65.9168

value of the parameter a_2 defines the relative height (h/H) at which the stem shape changes from a neiloid to a quadratic paraboloid, whilst a_1 defines the relative height above which it becomes conical. The parameter values determined for the four species are shown in Table 6.2.

Figure 6.2 shows the stem profiles predicted using this function for a tree of a particular diameter and height of each of the four species. It appears that *Pinus brutia* and *Cedrus libani* stems were rather similar in shape, whilst *Pinus taeda* had a rather more pronounced butt swell and *Abies cilicica* rather less so. Note also that the stem shape defined for *Pinus taeda* appears rather more complex than it does in Fig. 6.1, where the simpler function (6.6) was used. More complex versions of this functional form have been used successfully for other species (Clark et al. 1991; Özçelik et al. 2014).

Research continues in the development of taper functions of this nature. Many have now been developed for different species around the world, often using functional forms different from the examples given here (Hayward 1987; Bi 2000; Bi and Long 2001; Valentine and Gregoire 2001; Garber and Maguire 2003; Ter-Mikaelian et al. 2004; Jiang et al. 2005; Dean and Roxburgh 2006; Diéguez-Aranda et al. 2006b; Koskela et al. 2006; Lappi 2006; Zakrzewski and MacFarlane 2006; Newton and Sharma 2008; Li and Weiskittel 2010; Westfall and Scott 2010; Fonweban et al. 2011; Li et al. 2012; Goméz-García et al. 2013; Návar et al. 2013; Rodríguez et al. 2013; Ung et al. 2013).

Taper functions have been developed also to estimate not only stem wood volume but also the wood volume of the branches attached to the stem (Zakrzewski 2011; Ver Planck and MacFarlane 2014). Zakrzewski and Duchesne (2012) developed functions for jack pine (*Pinus banksiana*) in Canada that predicted how both stem diameter and wood density varied along the stem; use of these functions together allowed determination of how stem biomass varied along the stem.

Other research has developed functions to predict stem taper not just from tree diameter at breast height over bark and tree total height. Ung et al. (2013) devised functions for a number of Canadian tree species that required measurement of diameter at breast height only. Several authors found that functions requiring measurement of stem diameter at two different heights along the stem could improve estimates of stem diameter and wood volume (Goodwin 2009; Sharma and Parton 2009; Cao and Wang 2011); in effect, the two diameter measurements quantified the basic degree of taper of the stem. Özçelik et al. (2014) found that

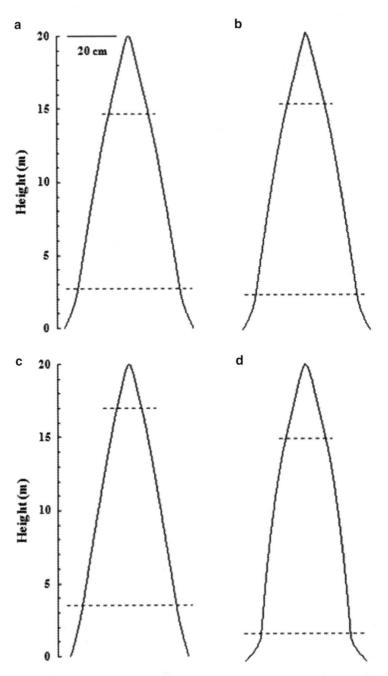

Fig. 6.2 Profiles of the stems of trees of (**a**) *Pinus brutia*, (**b**) *Cedrus libani*, (**c**) *Abies cilicica* and (**d**) *Pinus taeda*, determined using (6.8) with parameter estimates as given in Table 6.2. Each tree was assumed to have a diameter at breast height over bark of 45 cm and a height of 20 m. Below the *lower dashed line*, the stem shape is neiloidal; above the *upper dashed line*, it is conical. Between these points, it is shaped as a quadratic paraboloid (drawn using taper equations of Brooks et al. 2008 [**a–c**] and Sharma and Burkhart 2003 [**d**])

incorporating measurements of the length of the tree crown (vertical distance from the tree tip to the lowest green branch) and crown ratio (crown length divided by tree height) improved their taper function for Brutian pine (*Pinus brutia*) in Turkey.

Work has examined also how environmental and biological factors can affect taper functions. Crous et al. (2009) found that fertilization did not change the form of the taper function they developed for *Pinus patula* trees in an experimental plantation in South Africa. Nigh and Smith (2012) found that rainfall and frost incidence did change the taper function they developed for lodgepole pine (*Pinus contorta*) in Canada. Morley and Little (2012) found that taper functions for hybrid eucalypts in South African plantations differed both with genetic differences between individual trees and whether or not trees had single or double stems arising from cut stumps.

6.3.2 Using Taper Functions

The function developed for Tanzanian *Pinus patula* by Malimbwi and Philip (1989), (6.7), will be used as an example to illustrate how taper functions are used. For the sake of the example, suppose the only merchantable log products that can be sold from these plantations are **sawlog**s (logs large enough to be sawn into one or more of the many types of sawn wood used for building and many other purposes) or pulplogs. Suppose sawlogs are 2.4 m long and their under bark diameter at their small end must be no less than 15 cm. Suppose pulplogs are 3 m long and their under bark diameter at their small end must be no less than 8 cm.

Figure 6.3 shows the stem profile, predicted using (6.7), of the same *Pinus patula* tree as shown in Fig. 6.1, that is, with a diameter at breast height over bark of 23 cm and a total height of 21 m. The taper function may now be used to determine what sawlogs and pulplogs could be cut from this tree. Since sawlogs must have a diameter under bark at their small end of at least 15 cm, the first step is to determine how far up the stem it is before the diameter becomes less than this. This can be determined by rearranging algebraically (6.7) to give h on the left-hand side, that is,

$$h = H - \left[d_{\mathrm{uh}} H^{0.448} / \left(0.774 D^{0.932} \right) \right]^{(1/0.610)}. \tag{6.9}$$

Using (6.9), with $H = 21$ m, $D = 23$ cm and $d_{uh} = 15$ cm, shows that the stem diameter becomes 15 cm at 11.0 m from the base. This means that, at most, four sawlogs can be cut from the stem, since each must be 2.4 m long. Of course, when the tree is felled, it will be cut some little distance above its base, usually at a height (called the stump height) of about 0.2 m above ground. Thus, the four 2.4 m sawlogs could be cut from stem sections that are positioned 0.2–2.6, 2.6–5.0, 5.0–7.4 and 7.4–9.8 m along the stem. Whilst the last part of this section of the stem, 9.8–11.0 m, has a diameter large enough to be a sawlog, it is not long enough to be so. The positions of those four sawlogs are shown in Fig. 6.3.

Fig. 6.3 Under bark profile
of the stem of a *Pinus patula*
tree with a diameter at
breast height over bark of
23 cm and a height of 21 m.
For the example described
in the text, the *dashed lines*
show the positions from
which sawlogs (Saw) and
pulplogs (Pulp) could be cut
from the stem and what
would be wasted (Waste)
(derived using the taper
function of Malimbwi and
Philip 1989)

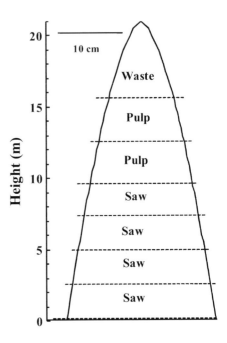

With four sawlogs cut from up to 9.8 m along the stem, the number of pulplogs that can be cut from the remainder of the stem can now be calculated. Using (6.9) with $d_{uh} = 8$ cm, the smallest diameter that the small end of a pulplog may have, it is found that pulplogs cannot be cut above 17.4 m along the stem. Thus, pulplogs could be cut from the 7.6 m long section extending over 9.8–17.4 m along the stem. Because pulplogs are 3 m long, two could be cut, positioned at 9.8–12.8 and 12.8–15.8 m along the stem. Their positions are marked also in Fig. 6.3. The remaining 5.2 m of the stem, 15.8–21 m, would be wasted.

In practice, the logs that could be cut from the stem will not occupy exactly the positions determined in this example. The few millimetres width of wood lost in cross-cutting with a chainsaw would have to be allowed for in positioning the logs. Also, where there are defects in the stem, a large branch swelling say, that point would be avoided when the stem is cross-cut. Sometimes, forest planners develop functions additional to their taper function to estimate where such defects are likely to occur on the stem and improve their estimates of exactly what logs can be cut from a particular stem.

Once it has been determined from what positions along the stem logs can be cut, the taper function allows additional information to be obtained about the size of each log. Knowing the heights in the stem at which the cross-cuts are made, (6.7) can be used to determine the diameters of each log at both its small and large end. Table 6.3 lists those values for the example tree.

Furthermore, the taper function can be used to determine the volume of each of the logs. To do this requires the use of integral calculus. Calculus in general (of which integral calculus is a part) is an extremely important and widely used

Table 6.3 Positions in the stem and diameters and volumes of the various logs, the stump and the waste section that would be cut from the example tree shown in Fig. 6.3. Results are shown also for the entire stem

Log	Position along stem (m)	Diameter		Volume (m³)
		Large end (cm)	Small end (cm)	
Stump	0–0.2	23.6	23.4	0.0087
Sawlog 1	0.2–2.6	23.4	21.7	0.0962
Sawlog 2	2.6–5.0	21.7	20.0	0.0820
Sawlog 3	5.0–7.4	20.0	18.1	0.0683
Sawlog 4	7.4–9.8	18.1	16.1	0.0550
Pulplog 1	9.8–12.8	16.1	13.3	0.0510
Pulplog 2	12.8–15.8	13.3	10.1	0.0325
Waste	15.8–21	10.1	0	0.0186
Entire stem	0–21	23.6	0	0.4123

mathematical tool invented in the late seventeenth century independently by an Englishman, Sir Isaac Newton, and a German, Gottfried Leibniz. This is not the place to discuss calculus; it is a complex field of mathematical study in its own right.

However, a brief idea can be given of how integral calculus would be used, with a taper function, to calculate the volume of a log positioned in a stem between any lower height above ground L (m) and an upper height U (m). Integral calculus first imagines that a very thin disk is cut from the stem at the lower height, L. The taper function is used to determine the diameter of the disk at that height and, hence, its circular cross-sectional area. The volume of the thin disk is then calculated assuming it is cylindrical in shape.

Integral calculus then imagines that a second thin disk is cut immediately above the first, and calculates its volume also; the second thin disk has a slightly smaller diameter than the first one because it comes from further up the stem. This process continues until the volumes of all the thin disks have been calculated right up to the upper height in the stem, U. Their volumes are then added up to give the total volume of the stem section between L and U. The beauty of integral calculus is that it is able to imagine that each thin disk is actually infinitesimally thin (i.e. it has no thickness at all). Thus, there is no error made by assuming that each disk is actually cylindrical, whereas a thin, but finite, disk would have a slightly smaller diameter at its upper end than at its lower end.

In the formal language of integral calculus, we write this process for the determination of the volume of a log (V_{LU}, m³) between heights L and U (m) in a stem as

$$V_{LU} = \int_L^U (\pi/40,000)d_h{}^2 dh. \tag{6.10}$$

In this relationship, the term $(\pi/40,000)d_h{}^2$ represents the circular cross-sectional area (m^2) of the stem at height h (m), where its diameter is d_h (cm), and the term dh (that is read in calculus as 'with respect to changing height along the stem') represents the (infinitesimally small) thickness (m) of the disk cut at that point. Their product $(\pi/40,000)d_h{}^2 dh$ is the volume (m^3) of the thin disk, assuming it is cylindrical. The terms L and U in (6.10) indicate that disks are to be cut between those two heights in the stem, and the special integral calculus symbol $\int$ (that is read as 'the integral of') means that the volumes of all the thin disks between L and U are to be summed to give their total volume V_{LU}.

Returning now to the example, substitute the right-hand side of the taper function (6.7) for d_h in (6.10) to give (with the squaring of d_h)

$$V_{LU} = \int_L^U (\pi/40,000)\left[0.599D^{1.864}(H-h)^{1.220}/H^{0.896}\right]dh. \tag{6.11}$$

It is now possible to rewrite this integral in a normal mathematical form that allows the volume to be calculated directly (there are mathematical textbooks on calculus that help you to do this integration). Doing so gives

$$V_{LU} = (\pi/40,000)\left[0.599D^{1.864}/H^{0.896}\right]\left[(H-L)^{2.220} - (H-U)^{2.220}\right]\Big/2.220. \tag{6.12}$$

Given this, consider the first sawlog cut from the base of the stem of the example tree (Fig. 6.3). It was positioned between heights 0.2 and 2.6 m along the stem, so for it $L = 0.2$ m and $U = 2.6$ m. Remembering that $D = 23$ cm and $H = 21$ m, (6.12) can be used to determine the volume of the log as 0.0962 m^3. Similarly, the volumes of each of the logs to be cut from the tree (and that of the stump at the base and waste section at the tip) can all be calculated using (6.12). The results are given in Table 6.3. Also, with $L = 0$ and $U = 21$, (6.12) can be used to determine the entire stem volume as 0.4123 m^3; of course, you will find that adding together the volumes of the various logs and those of the stump and waste will give exactly the same volume as this total.

It is important to note that it is not always possible algebraically to rearrange a taper function in order to use it directly to determine the height along the stem at which a particular diameter occurs (as was done with 6.9). Nor is it possible always to find the integral of a function as was done in obtaining (6.12) from (6.11). When no integral can be found, a simple method to determine approximately the volume of a log between any two heights along the stem is as follows. The taper function may be used to estimate the stem diameter at short intervals (say, 1 cm) along the stem section, Smalian's formula then used to estimate the section volumes that are then added together to give the log volume; this is equivalent to measuring the log sectionally. There are more sophisticated mathematical techniques available, called numerical techniques, that allow heights along the stem to be determined and volumes to be calculated when these mathematical difficulties are encountered; they are described in mathematics textbooks and will not be discussed further here. All these techniques involve considerable arithmetical computation that can be done practically only with the aid of a computer.

Given the discussion in this section, it can be seen that taper functions can be used to do all the things that stem volume functions can do. As well, they can be used to give additional detailed information about the logs that can be cut from a tree, information that is not available from stem volume functions. These days, most researchers prefer to develop taper functions rather than stem volume functions.

Chapter 7
Biomass

Abstract The biomass (weight) of trees is the most important measure of their growth. For climate change studies, it also measures the amount of the greenhouse gas carbon dioxide that has been taken up by trees from the atmosphere and, thus, how much carbon forests store. Measuring the biomass of large forest trees is a substantial undertaking, even more so if the root system is to be excavated and its biomass determined also. The methods used to fell, dissect and weigh tree leaf, branch, stem, coarse (larger, woody) roots and fine (small, living) roots are described. Work done throughout the world to develop individual tree biomass estimation functions from direct tree biomass measurements is then described. These allow estimation of tree biomass from simple measurements such as stem diameter at breast height, tree height and stem wood density. The application of these functions over local and continental scales is discussed.

7.1 Reasons for Biomass Measurement

Sale of logs for milling or papermaking has long been the principal market for the forestry industry and remains so. Because it is the volume of a log that determines if it is large enough to be sawn to produce timber, forestry has been concerned traditionally with measurement of stem wood volume. That is why two complete chapters (Chaps. 5 and 6) have been devoted to the topic in this book.

However, there is an increasing interest in measurement of the biomass, that is, the weight of trees. Furthermore, it is not only the stem biomass that is of interest but also the biomasses of the other parts of the tree, its leaves, branches and roots. Reasons for this interest include:

- All plants (including trees) produce biomass through photosynthesis. If scientists are to understand properly how tree growth occurs, they need to know how much biomass trees produce.
- Concerns with global warming. There is much interest in how much carbon is sequestered in forest biomass around the world as a result of trees taking up the greenhouse gas carbon dioxide.

© Springer International Publishing Switzerland 2015
P.W. West, *Tree and Forest Measurement*, 3rd edition,
DOI 10.1007/978-3-319-14708-6_7

- The establishment of plantation forests, to be grown for 3–5 years for bioenergy production. This wood is sold by weight, not volume.
- Products such as firewood or wood for papermaking are often sold by weight.

This chapter describes the techniques used to measure the biomass of various parts of individual trees. It then describes functions that have been developed to estimate tree biomass from easily measured tree characteristics.

7.2   Measuring Biomass

Felling, dissecting and weighing trees can be a major undertaking, even more so if the root system is excavated. A large forest tree, say with a diameter at breast height over bark of 35–40 cm and total height of 30–35 m, might weigh 3–5 tonne when freshly cut. Of this, 2–3 tonne might be the stem, 0.5–1 tonne might be roots and 0.3–0.5 tonne might be leaves and branches. Even bigger trees than this occur in forests and their total weight can exceed 20 tonne. By contrast, very young trees may weigh only a few kilograms, so it is trivial by comparison to fell and weigh them. Even then, excavation of the root system can be a tedious task.

Usually, it is desired to measure the dry biomass of the tree, that is, its weight after the water has been removed from the fresh biomass. Water makes up about 50 % of the weight of plants. Unlike nearly all the other biomass of a plant, water is not manufactured by the plant through photosynthesis or other metabolic processes. It is taken up from the soil by the roots. Thus, if the objective of the measurement is to determine what the plant has produced through its metabolic processes, it is the dry biomass that is relevant to measure, not the fresh biomass. Also, the amounts of water in plants can vary quite widely from time to time during the day or in different seasons. Thus, if consistency is required between measurements made of different trees at different times of year, dry biomass will be a more appropriate measure then fresh biomass.

To determine the **oven-dry** weight of biomass, fresh plant material is put in large laboratory ovens and dried at 60–80 °C for several days, until the weight of the dried material remains constant. Laboratory ovens are limited in size, so it is obviously impractical that all of a large forest tree should be dried; some form of sampling has to be done and only the sampled material is dried.

This section describes methods to measure the biomass of large trees. The difficulty of the various procedures described will decline steadily as the size of the tree declines.

7.2.1   Branches and Foliage

There are two strategies used to reduce the amount of work required to measure the biomass of branches and foliage after a tree is felled. The first involves counting all the branches that emerge from the main stem and then selecting a sample from amongst them, usually by a random selection procedure. Mathematically formal procedures for selecting the sample have been developed (Valentine et al. 1984; Gregoire et al. 1995; Temesgen et al. 2011). Leaves are removed (usually by tedious cutting with scissors) from the sample branches, and the branches themselves are cut into convenient lengths. The sample material is taken to the laboratory for drying. Sometimes also, the fresh weights of the sample leaves and branches are measured, by weighing with a large balance in the field; a randomly selected subsample is then weighed fresh in the field and only this material is dried.

Given the dry weight data from the sample, and knowing the number of branches included in the sample and on the whole tree, it is straightforward to estimate the total dry weight of all the branches and foliage in the crown. Sometimes it is desired to estimate also the branch bark separately from the branch wood. This requires that the bark be stripped from the sample branches and its weight determined separately. Bark removal can be very difficult, particularly at times of the year when the tree is not actively growing.

The second, and probably more precise, strategy for estimating branch and foliage weight is founded on the principle that branches must be sufficiently large to support the weight of the leaves to some distance out from the stem and to resist the forces imposed on them by the wind. Because of this, there is usually a close relationship between both branch weight and the weight of the foliage they support and easily measured branch characteristics, such as the diameter at their base and/or their length.

Given this, the diameter at the point where they join the stem of all the branches on a felled tree would be measured, together with their length. Where large branches have smaller branches arising from them along their length, the length is measured of the main branch only. A sample of branches from the tree is then selected and their foliage and branch material removed to the laboratory for drying. Using the sample data, regression relationships (Sect. 6.2.1) are established to allow estimation of foliage and branch dry weights, from branch diameter and/or length, for all the other branches along the stem that were not included in the sample. This process is a form of model-based sampling, a sampling method described in Sect. 10.5.

Often it is desired to measure the area of leaves, rather than their biomass. It is the leaf surface that absorbs light from the sun to provide the energy for photosynthesis and from which water is released by the leaves to the atmosphere. Thus, scientists studying plant growth behaviour often need to know leaf area instead of, or as well as, leaf biomass. The area of samples of leaves obtained for biomass measurement may be determined by placing them on a digital scanner; computer programs are then available to determine the area of the scanned image. In essence, this means that leaf area is being defined as the area of the shadow that a leaf casts when held horizontally over a flat surface, with light shining vertically down on it. The same definition is used for both broad- and needle-shaped leaves. Pérez-Harguindeguy et al. (2013) discussed leaf area measurement in more detail.

Often, both the leaf area and leaf oven-dry biomass are determined and the two values used to calculate the specific leaf area, that is, the area of the leaves per unit of their oven-dry biomass (its inverse, called leaf mass per area or specific leaf weight is often reported also). In essence, specific leaf area is a measure of the thickness of the leaves. It varies widely, perhaps in the range 2–40 m^2/kg, with tree species and tends to be lower in older trees (Holdaway et al. 2008), in trees growing in wetter environments (Gouveia and Freitas 2008), in overstorey plants growing in full sunlight (e.g. Specht and Specht 1999; Atwell et al. 1999) and in the better lit, upper parts of the **canopy** (the foliage and branches of the forest) (Monserud and Marshall 1999; Grote and Reiter 2004).

Dead branches present a problem in biomass measurement. They are usually found near the base of the crown where shading from leaves higher up has led to loss of leaves from the branch. Eventually dead branches are shed by the tree, but if they are still attached to the tree at the time of measurement, it is usually considered appropriate to include them as part of the tree biomass. Such branches, and indeed branches with only a few leaves left, may have to be measured and sampled separately from other branches.

Often, there are problems near the tip of the tree. For many species, it is difficult to identify what constitutes the main stem where many small, upright branches are growing near the tip. It may be necessary to treat the tip region separately, cutting it off and weighing all its leaves and branches, perhaps taking a subsample only away for drying.

7.2.2 Stems

Biomass of stems is usually determined by measuring stem volume by sectional measurement (Sect. 5.3). To convert volume to biomass, stem wood density must then be determined also.

Stem wood density varies both vertically along the length of tree stems and horizontally over their cross section at any point along the stem (West 2014a, Sect. 3.3.3). For biomass estimation, some average density for the whole stem is required that may then be multiplied by stem volume to convert it to biomass. Usually, a sample of stem discs, each a few centimetres thick, is cut from the stem at varying distances along it. Mathematically formal sampling strategies are sometimes used to determine the points along the stem at which the discs are taken (Valentine et al. 1984). In the laboratory, the volume of each disc is measured (by water immersion or by measuring its diameter and thickness). The disc is then oven-dried and weighed. This gives the **basic density** of wood, defined as the oven-dry weight of wood per unit fresh volume. In this book units of kg/m^3 are used for basic density, although other works often use units of g/cm^3; multiplication of the latter by 1,000 converts them to the former units. The average over all the discs is used as an average for the whole stem. Williamson and Wiemann (2010) and Pérez-Harguindeguy et al. (2013) discussed various issues that need to be considered in measuring wood density. Note that basic specific gravity (or sometimes just specific gravity) of wood is often referred to in the literature rather than basic density. This is simply basic density as a proportion of the density of water at 4.4 °C, which is 1,000 kg/m^3. Thus, basic specific gravity varies over the range 0–1 and is dimensionless (Williamson and Wiemann 2010).

Destructive sampling of trees to measure stem wood density in this way is difficult and time consuming. It would obviously be desirable if measurements could be made on standing trees. A method to do so uses devices that may be screwed into the tree stem and a core of wood removed from it (Fig. 7.1). These devices are used regularly by scientists who study climate change by measuring the width of growth rings in tree stems, a field of study known as dendrochronology. They may cause some damage to trees of some species in some circumstances, but generally do little harm (Eckstein and Dujesiefken 1999; Grissino-Mayer 2003; van Mantgem and Stephenson 2004; Wunder et al. 2011).

Wiemann and Williamson (2012) found that it may be possible to estimate the average density across the whole cross-sectional area at a particular point along a stem by removing a core to a depth of only one sixth of the stem diameter; this would reduce the time and effort required to obtain a core from any tree. They suggested that wood density measured only from the lower part of the stem, at breast height say, may be adequate for some purposes for which wood density is used; this might be so for uses such as biomass estimation functions that will be discussed in Sect. 7.3. However, since stem wood density varies both vertically and

Fig. 7.1 A device used to extract a wood core from the stem of a tree. The main part of the instrument (the T-section) is screwed into the tree stem to the depth desired. The long secondary device shown is then inserted down the centre of the core tube to facilitate the removal of the core from the stem. These devices are available in many different sizes to deal with trees of different sizes (Image provided by courtesy of Forestry Tools Australia, http://www.forestrytools.com.au)

horizontally, measurement high up on the stem may be necessary if an unbiased estimate of average stem wood density for a whole tree stem is required.

Chave et al. (2006) collated wood density data available from 2,456 tree species from Central and South America. Their basic densities ranged over 110–1,390 kg/m^3 with an average of 645 kg/m^3. They found that wood density tended to be lower in species growing at higher altitudes. In forests growing at lower altitudes, it differed both with rainfall and geographic location. They found also that knowledge of the genus to which a species belonged was often an adequate guide as to what its wood density was likely to be. For trees from rainforests in Ghana, Africa, Henry et al. (2010) found that wood density of species that grow rapidly in full sunlight after a disturbance that opens the forest canopy (such as a storm or death of a large tree that falls) averaged 420 kg/m^3, whilst that of slower-growing species that develop subsequently averaged 630–640 kg/m^3: this process of forest 'succession' is described further in Sect. 8.3.

7.2.3 Roots

Root biomass measurement is vastly more difficult than measurement of the above-ground tree parts. The problems include the difficulties of physical excavation of roots from the soil and the inability to identify whether a root belongs to the tree being measured, or is part of the overlapping root system of a neighbouring tree or is a root from an understorey species. These difficulties can never be solved completely. Thus, it is inevitable that root biomass measurements will tend to be less precise than above-ground biomass measurements.

One method of measuring root biomass is to undertake a full excavation of the soil around the stump of a tree. A trench might be dug around the stump with a backhoe, at a distance (perhaps about 1–2 m) from the tree and to a depth (perhaps about 1 m) within which it is judged most of the roots of the tree are located. Hand tools and, perhaps, water or air pressure equipment or sieves would then be used to manually sift through the soil, finding and extracting the roots; obviously, this is a very labour-intensive and tedious task (e.g. Di Iorio et al. 2005; Peichl and Arain 2007; Borden et al. 2014; Herrero et al. 2014).

The work involved in root excavation can be reduced by taking soil core samples around the stump. Usually, the stump itself will have been pulled out with a machine, removing with it as many large roots as possible. Cores may then be taken, with a hand or machine auger, to sample the distance and depth within which it is believed the roots will lie. The cores are usually taken to the laboratory to sort the roots from the soil.

It is often desired to distinguish between coarse and fine roots. Coarse roots include the large, strong, woody roots that extend immediately out from the base of the tree and anchor it firmly in the ground. An extensive web of smaller coarse roots ramifies from these larger roots. Coarse roots act also as part of the transport system of water and nutrients through the tree. Water and nutrients are taken up from the soil by fine roots and pass through the wood of the coarse roots, the stem and the branches up to the leaves. Wood consists mainly of dead tissue, and water, with nutrients dissolved in it, can pass through the empty, dead wood cells; these are joined to each other by holes to make a system that can be thought of as hollow pipes right up the whole length of the tree.

Fine roots occur at the extremities of the coarse root system and consist of live tissue capable of taking in water and nutrients from the soil. They have a limited lifespan that can be as short as a few months. When they die, they become part of the organic matter of the soil and are eventually broken down by bacteria and fungi, so the nutrients they contain become available again in the soil for reuse by the tree. New fine roots then develop in their place.

When excavating roots, it is very difficult to determine exactly where, along its length, a fine root starts to develop wood and become a coarse root. In root biomass measurement, it is common to make the rather arbitrary decision that fine roots are those less than about 2 mm in diameter and anything larger is considered to be a coarse root. As well, it is often difficult to tell whether a fine root is still alive or has died; often its colour and its strength when pulled can be used to distinguish live from dead root tissue (Vogt et al. 1998).

It is often desired to measure both the biomass and the turnover rate of fine roots without felling a tree and excavating its root system. The three main techniques used to do so are (Vogt et al. 1998):

- Soil coring—if the fine root biomass is measured in a series of soil cores taken at several times during a year, their turnover rate can be estimated. Lee et al. (2004) give an interesting example of the use of soil cores to produce a map, at quite fine scale, of the fine root biomass distribution around individual trees in conifer forest in north-western USA.
- Ingrowth cores—a mesh sleeve, containing root free soil, is inserted into the hole left by removing a soil core. Periodic measurement of the biomass of live and dead fine roots that appear in such sleeves, as roots grow into them, gives estimates of both their biomass and turnover rate.
- Minirhizotrons—these are transparent tubes inserted into the ground. Using mini-cameras or other electronic viewing devices to look down the tubes, recordings are made of fine roots as they grow around the outside of the tube and are visible at its surface. Flat glass plates inserted in the soil may be used also and the growth of roots along the glass surface recorded. Coupled with biomass measurements made from cores, this provides information from which fine root turnover rates may be determined.

Many refinements of these techniques of both coarse and fine root measurement have been tested from time to time for various species in many different soil circumstances. Works such as Vogt et al. (1998), Bengough et al. (2000), Snowdon et al. (2002) and Danjon and Reubens (2008) should be consulted before root biomass measurements are attempted. So difficult is root measurement, a substantial proportion may be missed; this has often led to underestimation of root biomass by perhaps as much as 30–40 % (Robinson 2004) or to otherwise inadequate estimates (Mokany et al. 2006).

7.2.4 Carbon Content of Biomass

As mentioned earlier (Sects. 1.2 and 7.1), concerns about global warming have led to considerable interest in determining how much of plant biomass is carbon. Direct measurement of the amount of carbon in plant biomass is a specialist laboratory process. It involves grinding samples of the dried biomass to a fine powder, burning the samples and measuring the amount of carbon dioxide given off using a complex laboratory instrument known as a mass spectrometer that 'weighs' atoms or molecules. Knowing the weight of the original sample, the proportion of it that was carbon can then be determined.

Thomas and Martin (2012) collated data from the literature and obtained information for 253 tree species scattered widely around the world. On average, the oven-dry stem wood of hardwood species contained 48 % carbon, whilst that of softwood species averaged 51 %. Over all species, the range of carbon contents was 40–60 %. Further, they found that the carbon contents of other tissues, leaves, twigs, branches and coarse and fine roots were generally similar to that of stem wood.

It has often been assumed that 50 % of the oven-dry biomass of trees is carbon, and this figure has been used widely in the past when attempts have been made to estimate the amount of carbon stored in forest tissue. Thomas and Martin's work showed that whilst some species do have carbon contents that differ somewhat from this value, the overall average for tree species of the world lies close to it. However, they did point out that some carbon is lost from plant tissue when it is oven-dried and that actual carbon contents may be 1–2 % higher than those reported for oven-dry tissue.

Sometimes the amount of carbon stored in biomass is reported as an equivalent amount of carbon dioxide that the tree has removed from the atmosphere. The conversion from carbon to carbon dioxide equivalent is done simply by multiplying the carbon amount by 3.67; this is the ratio of the weight of a molecule of carbon dioxide to the weight of an atom of carbon.

7.3 Above-Ground Biomass Estimation Functions

Given the difficulties involved with direct measurement of tree biomass, it is not surprising that attempts have been made to develop functions to allow tree biomass estimation from simply measured characteristics of standing trees, such as their stem diameter and height. These functions fill the same role for biomass estimation as the functions discussed in Chap. 6 do for tree stem volume estimation. However, interest in tree stem volumes has been concerned generally with estimation of amounts of wood available for conversion to various forest products (Sect. 1.2). Interest in biomass has developed both as part of scientific study of tree growth behaviour and, over recent times particularly, in estimating the amount of carbon dioxide sequestered within tree biomass as part of climate change research (Sect. 1.2).

Perhaps most commonly, the basis of biomass estimation for a given species in a given part of the world has been a function that estimates the above-ground biomass of an individual tree (its stem plus branches and leaves). The functional form used most commonly for this predicts oven-dry, above-ground biomass of an individual tree (B_A, tonne) from its diameter at breast height over bark (D, cm) as

$$B_A = \alpha D^\beta, \tag{7.1}$$

where α and β are parameters. There are many examples of the use of this function for many different tree species around the world (Freedman 1984; Ter-Mikaelian and Korzukhin 1997; Eamus et al. 2000; Burrows et al. 2001; Saenger 2002; Jenkins et al. 2003; Pérez Cordero and Kanninen 2003; Specht and West 2003; Van Camp et al. 2004; Xiao and Ceulemans 2004a; Hamilton et al. 2005; Montagu et al. 2005; Kajimoto et al. 2006; O'Grady et al. 2006; Wang 2006; Zerihun et al. 2006; Johansson 2007; Liddell et al. 2007; Rock 2007; Case and Hall 2008; Paul et al. 2008; Ung et al. 2008; Kenzo et al. 2009b; Annighöfer et al. 2012; Mate et al. 2014). Even in cases of multi-stemmed trees, when each stem is considered to be a separate tree, (7.1) has been found effective (Hamilton et al. 2005). It is a simple form of (6.1) that was discussed earlier as a function used to estimate tree stem wood volume.

Functions such as (7.1) that relate plant biomass to one or more other variables that reflect the size of the plant (such as its stem diameter or height) are often called allometric relationships. The term **allometry** means the relationship between part of an organism and its whole. Several authors (Causton 1985; Parresol 1999; Verwijst and Telenius 1999) have reviewed the use of allometric functions for biomass estimation. Considerable care is required in applying regression analysis (Sect. 6.2.1) to obtain values for the parameters of allometric functions (Bi et al. 2001; Lai et al. 2013; Mascaro et al. 2014).

Just as has been found with (6.1) for tree stem wood volume, researchers working with various species in various parts of the world have used functional forms other than (7.1) that sometimes include tree height rather than diameter alone (Schmitt and Grigal 1981; Kumar and Tewari 1999; Saenger 2002; Carvalho and Parresol 2003; Pitt and Bell 2004; Zerihun and Montagu 2004; Khan et al. 2005; Saint-André et al. 2005; Brandeis et al. 2006; Cole and Ewel 2006; Cienciala et al. 2006; Williams and Gresham 2006; Dillen et al. 2007; Peichl and Arain 2007; Pajtík et al. 2008; Paul et al. 2008; Ung et al. 2008; Kenzo et al. 2009b); Mason et al. (2014) applied such a function to woody shrubs, rather than trees, in New Zealand. However, sometimes there has been found to be little advantage from the inclusion of height in the function (Ter-Mikaelian and Korzukhin 1997;

Burrows et al. 2001; Kajimoto et al. 2006; Wang 2006) and there may even be a disadvantage (Montagu et al. 2005).

Occasionally functions have been used in which diameter is the only prediction variable but that have a form different from that of (7.1) (Droppelmann and Berliner 2000; Chambers et al. 2001; Montagu et al. 2005; Ximenes et al. 2006; Dillen et al. 2007). Sometimes other tree characteristics or variables that reflect environmental differences between different sites have been included in the function (Droppelmann and Berliner 2000; Carvalho and Parresol 2003; Pérez Cordero and Kanninen 2003; Xiao and Ceulemans 2004a; Hamilton et al. 2005; Saint-André et al. 2005; Cienciala et al. 2006; Cole and Ewel 2006; Peichl and Arain 2007; Paul et al. 2008; Goodman et al. 2014; Zell et al. 2014).

The biomass estimation functions discussed above have often been developed for local use with individual species in particular parts of the world. However, especially because of interest in the amounts of carbon stored in forests as part of climate change research (Sects. 1.2 and 7.1), there have been attempts to develop biomass estimation functions that apply widely to trees of many species, of widely ranging sizes and ages, and that are spread over a geographical region of considerable size and/or over a range of environmental circumstances. If such functions can be developed, they will go some way to reducing the substantial work involved in direct measurement of tree biomass (Sect. 7.2) necessary to develop a function for any one species growing in particular circumstances.

A first example of such functions comes from Chojnacky et al. (2014) who updated the work of Jenkins et al. (2003). They summarised results from many North American studies, all of which used (7.1), for a wide range of both hardwood and softwood species from many different locations across the USA and Canada. They identified 35 taxonomic groupings of species in each of which they felt a common biomass function might apply and that might be used reliably for the members of each group right across North America. Similar work has been done for other species in other parts of the world (Muukkonen 2007; Rock 2007; Ung et al. 2008).

A second example comes from Williams et al. (2005). They were concerned with estimating the above-ground biomass of trees of a variety of tree species (mainly eucalypts) across the vast areas of **woodland** forests that cover tropical and subtropical northern Australia. They had measured tree biomasses at 20 geographically widespread locations in these forests. The best function they found for their purpose included both tree diameter at breast height and tree height. It was

$$B_\mathrm{A} = 0.0001275 D^{2.1735} H^{0.1362\ln(H)}, \tag{7.2}$$

where H was tree height (m) and $\ln(H)$ denotes the natural logarithm of H. This function seemed to predict above-ground biomasses adequately for any tree of the many species at many sites for which they had data.

Other authors have tried to develop functions that might apply across even larger scales and to a wider range of species than those described above. In doing so, they have obtained the best results by using functions that include stem wood basic density (Sect. 7.2.2) as one of the predicting variables. The inclusion of basic density reflects the fact that it relates to the strength of wood (Chave et al. 2009; West 2014a, Sect. 3.3.3). Thus, the height, diameter and taper of a tree stem will interact with its wood density to ensure the tree has sufficient biomechanical integrity to remain standing; the effects of wind forces on stem taper were discussed in Sect. 6.2.2.

To develop a function of this type, Pilli et al. (2006) collated above-ground biomass data from destructive harvest of 1,278 trees of 26 hardwood and 11 soft-wood species scattered widely across the northern hemisphere, mainly north of the tropical regions. They included stem wood basic density as part of their biomass function and found also that the position of the tree within the hierarchy of tree sizes in the forest was important in determining its above-ground biomass. Considering just the larger trees in a stand, their function was

$$B_A = \exp(-10.03 + 0.00111\rho)\, D^{2.51}, \tag{7.3}$$

where ρ (kg/m^3) was wood basic density averaged over the whole stem of the tree. Note that the meaning of the mathematical term 'exp' (which is an abbreviation of 'exponential') is explained in the footnote to Table 6.1.

A further example of functions of this type was provided by Chave et al. (2005), now updated by Chave et al. (2014) after discussion by various authors (Kenzo et al. 2009a; Djomo et al. 2010; Henry et al. 2010; van Breugel et al. 2011; Alvarez et al. 2012; Feldpausch et al. 2012; Banin et al. 2014; Goodman et al. 2014; Mbow et al. 2014; Ngomanda et al. 2014). Chave et al. (2014) collated above-ground biomass data from destructive harvest of 4,004 trees at 53 undisturbed and 5 sec-ondary sites (sites where trees have regrown after clearing) distributed right across the tropical forests of Central and South America, Africa and Southeast Asia. The sites covered a wide range of temperature and rainfall environments from lowlands to highlands. The single function that fitted their data well, right across this wide range of forests, was (their equation (7.4), with adjustment for the units used here)

$$B_A = 7.95 \times 10^{-8} \left(\rho D^2 H\right)^{0.976}. \tag{7.4}$$

They developed also a function to use if tree height was not measured. However, it was found they could achieve a satisfactory result in that case only if they included an environmental variable in the function that reflected seasonal variation in temperature and rainfall as well as the incidence of drought at the site concerned.

Functions including wood basic density have been found useful at a smaller scale, but where a number of different species occur together. Thus, Colgan et al. (2014) were able to determine a single function that estimated individual tree stem biomasses satisfactorily across a number of species of the savanna woodlands of South Africa.

Given the large amount of work necessary to obtain biomass data by destructive harvest of trees (Sect. 7.2) to develop a biomass function for any particular species in any particular region, functions that apply across wide geographic areas and a variety of forest types seem very appealing. They have been put to use at these scales. For example, the functions of Chave et al. (2005) were used by Banin et al. (2014) in comparing growth rates of tropical forests of the Amazon Basin and of Borneo and by Cavanaugh et al. (2014) in comparing the biodiversity of tropical forests of the Americas, Africa and Asia and its ramifications for their ability to store carbon.

However, it is apparent from collations of biomass functions for many species (e.g. Henry et al. 2011; Chojnacky et al. 2014) how widely functions can differ between species or species groups. This means that substantial bias can result if the wide-ranging biomass estimation functions are used at a small scale, that is, to estimate biomasses of a few species only over a particular region or at particular sites. For example, Vieilledent et al. (2012) found that the wide-ranging tropical forest functions of Chave et al. (2005) tended to overestimate biomasses of larger trees, particularly when applied to forests at several sites in Madagascar; Chave et al. had no data available from Madagascar when they devised their functions. In the revision of their functions, Chave et al. (2014) themselves found that there were a number of regions where their function tended to overestimate tree biomasses substantially when compared with locally developed models. Other researchers have considered this problem also (Komiyama et al. 2008; Kenzo et al. 2009a; Návar 2009; Djomo et al. 2010; Henry et al. 2010; van Breugel et al. 2011; Alvarez et al. 2012; Goodman et al. 2014; Mbow et al. 2014; Ngomanda et al. 2014).

Thus, it seems that research has not yet discovered a single relationship between individual tree above-ground biomass and readily measured tree characteristics, such as stem diameter, tree height and stem wood basic density, that applies across all tree species right across the world. Certainly this is the case for tree stem wood volume functions (Chap. 6), and research has been exploring those for much longer than has been the case for biomass functions. Perhaps this is not surprising, given the enormous genetic variation between different tree taxa (the named units by which living organisms are classified, including species, genus etc.) across the world, variation that inevitably affects the allometric relationships between their parts. This means that it may be necessary for many years hence to continue to develop biomass estimation functions for different tree species, or at least species groups, if adequate, unbiased estimates of their biomasses are to be obtained readily.

To develop such functions, it will be necessary to undertake the arduous and expensive task of measuring the biomasses directly of a sample of felled trees of any species of interest in a region for which it is desired to develop a biomass estimation function. Research has been investigating sophisticated approaches that aim to reduce to a minimum the size of the direct measurement sample necessary to develop a function adequately for a new species or forest type (Zianis and Mencuccini 2004; Picard et al. 2012; Zapata-Cuartas et al. 2012).

7.4   Biomass Estimation Functions for Tree Parts

Development of functions to estimate above-ground biomass of individual trees (Sect. 7.3) has received rather more attention than functions to estimate biomasses of other tree parts (leaves, branches, stems, roots, etc.) for two reasons. Firstly, when obtaining data with which to build biomass estimation functions, it is far more common to measure above-ground biomass only because root biomass is so difficult to measure (Sect. 7.2). Secondly, research emphasis has been on estimation of the amounts of carbon sequestered by whole forests through photosynthesis and so has concentrated on functions that estimate total biomass, rather than biomass of the various tree parts.

Nevertheless, in scientific studies of tree growth behaviour, it is common to wish to know how the different parts of the tree develop with time. Thus, some attention has been paid to development of functions to estimate the biomass of the various parts of trees.

7.4.1   Allometric Functions

Allometric functions, of the same form as (7.1), have been used to estimate the biomasses of various parts of individual trees, with the biomass of the part concerned replacing B_A in the equation. Figure 7.2 shows a typical example, in this case for sugar gum (*Eucalyptus cladocalyx*) being grown in plantations in southeastern Australia (Paul et al. 2008). In this case, the biomass functions developed for each tree part seemed to predict biomass satisfactorily across the three plantation sites from which Paul et al. had collected data, where the plantation age varied over the range 5–75 years.

Paul et al. determined also a function to predict above-ground biomass in their plantations, using (7.1). For interest, this function and that for large trees of Pilli et al. (2006) are compared in Fig. 7.2. There seems reasonable agreement between Paul et al.'s and Pilli et al.'s functions; as discussed in Sect. 7.3, Pilli et al.'s function shows potential for general use for a wide range of tree species anywhere in the world.

Of course, if the biomasses of various parts of a tree have been estimated, they may be summed to give the total biomass of the tree. So, in Paul et al.'s case, the estimates for leaves, the stem and branches might be summed to give the above-ground biomass of the tree. When this is done, it is found that the predicted above-ground biomass does not agree with that estimated using the function they developed to predict the above-ground biomass directly. In Paul et al.'s case, the discrepancy is quite small. For example, for a 47 cm diameter tree, the above-ground biomass estimated by summing the estimates for the parts is 2.15 tonne, whilst that obtained directly from the above-ground biomass function is 2.12 tonne. However, such discrepancies can be larger, depending on the forest circumstances. They arise because none of the functions is an exact predictor of biomass; each

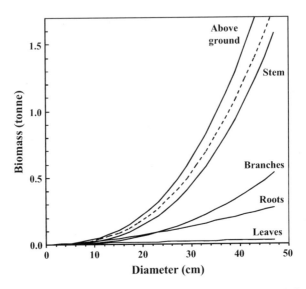

Fig. 7.2 For various parts of the tree as marked, the *solid lines* show the relationship between oven-dry biomass and tree diameter, as determined for *Eucalyptus cladocalyx* by Paul et al. (2008). The *dashed line* shows the relationship for above-ground biomass from the function of Pilli et al. (2006) for 'larger' trees (7.3); to obtain that line required the use also of the function determined by Paul et al. to estimate stem wood basic density of *E. cladocalyx* trees from their stem diameter

function gives an estimate only. Mathematical statistical techniques have been developed to avoid this problem and force estimates for individual parts to sum exactly to the value estimated using the function that predicts the total directly (Parresol 1999; Bi et al. 2001; Carvalho and Parresol 2003). These techniques only ensure that the sum of the estimates made for the parts equals the estimate made directly for the total: they still all remain estimates, albeit with consistency across all the functions involved.

There are many other examples where allometric functions have been developed to estimate biomasses of different parts of trees. These include cases where a function was determined from a sample of trees in one locality, for local use, or where it was attempted to develop functions that applied to a variety of species over wide geographical regions (Ter-Mikaelian and Korzukhin 1997; Droppelmann and Berliner 2000; Burrows et al. 2001; Saenger 2002; Jenkins et al. 2003; Pérez Cordero and Kanninen 2003; Pitt and Bell 2004; Xiao and Ceulemans 2004a; Zerihun and Montagu 2004; Kajimoto et al. 2006; Wang 2006; Zerihun et al. 2006; Muukkonen 2007; Coll et al. 2008; Coyle et al. 2008; Holdaway et al. 2008; Levia 2008; Paul et al. 2008; Sochacki et al. 2007; Ung et al. 2008; Wutzler et al. 2008; Návar 2009; Chojnacky et al. 2014; Mate et al. 2014). Some researchers have found, empirically, that inclusion of tree characteristics other than stem diameter (such as tree height or the length of the crown of the tree) has improved the function, whilst others have found diameter by itself was adequate

(Burrows et al. 2001; Carvalho and Parresol 2003; Pérez Cordero and Kanninen 2003; Pitt and Bell 2004; Xiao and Ceulemans 2004a; Zerihun and Montagu 2004; Khan et al. 2005; Brandeis et al. 2006; Cienciala et al. 2006; Cole and Ewel 2006; Massada et al. 2006; Tobin et al. 2006; Paul et al. 2008; Wutzler et al. 2008). Sometimes a functional form different from the allometric form of (7.1) has been used (Jonckheere et al. 2005a; Saint-André et al. 2005; Xing et al. 2005; Kajimoto et al. 2006; Williams and Gresham 2006; Dillen et al. 2007; Muukkonen 2007; Pajtík et al. 2008; Paul et al. 2008; Annighöfer et al. 2012; Herrero et al. 2014).

7.4.2   Biomass Expansion Factors

Over the second half of the twentieth century, many tree stem volume functions (Sect. 6.2) have been developed throughout the world for different tree species growing under a wide range of environmental circumstances. A second form of biomass estimation function for individual tree parts exploits the availability of these volume functions. In these cases, the ratio of the biomass of one or more parts of a tree to the stem volume of the tree is determined. Such a ratio is known as a **biomass expansion factor**; multiplication of stem volume, determined using the volume function, by the biomass expansion factor provides an estimate of biomass for the relevant part of a tree.

An example can be found in Peichl and Arain (2007) for **native forests** of white pine (*Pinus strobus*) in Canada. They measured biomasses of the various above-ground parts of trees in forests of varying age. They found that individual tree biomass expansion factors for leaves, branches and roots varied with forest age and developed empirical functions to relate those factors to age. The results are illus-trated in Fig. 7.3. Other examples can be found in Rytter (2006) and Pajtík et al. (2008).

7.4.3   Leaves

Whilst the methods described in Sects. 7.4.1 and 7.4.2 do offer functions to estimate tree leaf biomass, it is often found that the estimates they provide are not as good as might be desired. This reflects the biological circumstances of leaves in forests.

As a young tree grows over its first few years of life, its leaf biomass increases with age. However, eventually there reaches a point in the life of a forest when the leaf biomass over the forest as a whole reaches a more or less constant value. This value is determined principally by the availability to the trees of the water and nutrients from the soil at the site. Once this point is reached, stems, hence above-ground biomass, of individual trees will continue to increase in size with time whilst the biomass of their leaves will remain more or less steady. This situation will become even more complicated as the competitively more successful trees in a stand outgrow their less successful rivals. The leaf biomass of a less successful tree

Fig. 7.3 Biomass
expansion factors, as a
function of tree age, for
different parts of individual
trees of *Pinus strobus* in
forests of southern Ontario,
Canada. Multiplication of
tree stem wood volume (m³/
ha) by the biomass
expansion factor will give
an estimate of the biomass
of each part of the tree
(derived from information
in Table 6 of Peichl and
Arain 2007)

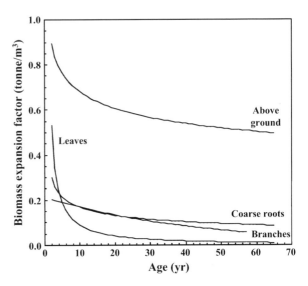

will then decline gradually with age until the tree dies, whilst the leaf biomass of a
more successful tree will increase. Examples of the effects of different competitive
pressures on tree foliage biomasses can be found in Baldwin et al. (2000) and
Holdaway et al. (2008).

Given this complexity, it is obvious that it will be difficult to build biomass
functions generally to predict individual tree leaf biomasses over the whole life of
the forest. At any one age, simple allometric equations (Sect. 7.4.1) have been used
successfully for this purpose, but it is to be expected that the parameter values of
these functions would change as the trees age and as their position in the stand
alters. This is not always so, as the example in Fig. 7.2 illustrates, where the
functions developed applied satisfactorily across a wide range of plantation ages.
Nevertheless, there are many cases where it is so. For example, Holdaway
et al. (2008) found variation with age in natural stands of mountain beech
(Nothofagus solandri var. *diffortioides)* in New Zealand. For Sitka spruce (*Picea
sitchensis*) in Ireland, where stand age varied over the range 10–46 years, Tobin
et al. (2006) needed to include all of tree stem diameter, the length of the crown and
the **stocking density** of the forest in which it occurred as variables in their biomass
estimation function. Grote and Reiter (2004) related the crown biomass of individual European beech (*Fagus sylvatica*) and Norway spruce (*Picea abies*) trees to the
competitive influence the trees of the forest exerted on each other. There are other
examples of leaf biomass functions that include various tree or stand parameters to
allow for changes in age or other environmental circumstances of the tree
(Monserud and Marshall 1999; Pérez Cordero and Kanninen 2003; Pitt and Bell
2004).

As mentioned in Sect. 7.2.1, it is often the area of leaves that is of interest, rather
than their biomass. Arias et al. (2007) and Macfarlane et al. (2007a) developed leaf

area estimation functions, using (7.1) with leaf area replacing biomass, for various species in Costa Rica and Australia, respectively.

A quite different tree variable, the cross-sectional area of sapwood in the stem, has been found particularly useful in developing leaf biomass functions. Water is transported from the roots up through the sapwood to the leaves. Sapwood occurs in the outer part of the stem cross section and consists of empty, dead cells through which water can pass. Heartwood occurs nearest the centre of the stem, and its cells are blocked by resins and other substances so water no longer can pass through it. Sapwood is continually converted to heartwood throughout the life of the tree. However, the area of the sapwood in the stem at any time must be sufficient to allow passage of the volume of water necessary to supply the needs of the leaves. Thus, it can be expected that there will be a close relationship between stem sapwood area and the biomass of the leaves carried on the tree.

The width of sapwood can be measured on a core taken from the stem of a standing tree (Sect. 7.2.2 and Fig. 7.1). Knowing the diameter of the stem, the sapwood area may then be determined. There are numerous examples of functions developed to predict tree foliage biomass from stem sapwood area (Whitehead et al. 1984; Pothier et al. 1989; West and Wells 1990; Shelburne et al. 1993; Pereira et al. 1997; Berninger and Nikinmaa 1998; Monserud and Marshall 1999; Medhurst and Beadle 2002; Gilmore and Seymour 2004; Stancioiu and O'Hara 2005; Xiao and Ceulemans 2004b; Kajimoto et al. 2006; Tobin et al. 2006).

7.4.4 Roots

There are numerous examples (one is illustrated in Fig. 7.2) where coarse or total (coarse plus fine) tree root biomass has been estimated from tree stem diameter at breast height using (7.1), with root biomass replacing above-ground biomass (Burrows et al. 2001; Kajimoto et al. 2006; Miller et al. 2006; Wang 2006; Zerihun et al. 2006; Johansson 2007; Lavigne and Krasowski 2007; Peichl and Arain 2007; Vadeboncoeur et al. 2007; Coll et al. 2008; Paul et al. 2008; Návar 2009; Brassard et al. 2011), or using other functions that include stem diameter only (Tatarinov et al. 2008; Herrero et al. 2014). Sometimes tree variables other than stem diameter have been included also (Cole and Ewel 2006; Sochacki et al. 2007; Paul et al. 2008; Herrero et al. 2014).

However, when roots are excavated in a forest area, it is difficult to identify to which individual tree any particular root belongs (Sect. 7.2.3). Perhaps because of this, there has been an emphasis in the development of root biomass functions that estimate the root biomass per unit ground area of the forest, rather than the root biomass of individual trees. This is the next scaleup of measurement to be considered in this book (Sect. 1.3); root biomass estimation functions of that nature will be discussed in Sect. 8.9.

Fine roots present rather more of a problem than coarse roots. Generally, fine roots constitute a small proportion of the total biomass of root systems, perhaps tending to decline with age to about 10 % of total root biomass as the forest matures. This can vary widely with different forests in different environmental circumstances. For example, Kajimoto et al. (2006) found that the fine root biomass in Dahurian larch (*Larix gmelinii*) forests, in the cold, boreal regions of Siberia, made up 58 % of the total root biomass in younger forests (trees 95–100 years old) and 34 % in older forest (trees up to 280 years old). However, the lifespan of fine roots can be short (as little as a few months, e.g. Jourdan et al. 2008; Gaul et al. 2008), so they turn over continuously as some die and are replaced by new ones.

Development of functions to estimate the biomass of fine roots of individual trees is far less advanced than for other tree parts. However, Helmisaari et al. (2007) developed such a function satisfactorily for trees of Norway spruce (*Picea abies*) and Scots pine (*Pinus sylvestris*) in native forest stands in Finland; they used (7.1) with fine root biomass replacing above-ground. However, they found that trees tended to have greater fine root biomasses on sites that were cooler or had less fertile soil; thus the functions they developed were site specific. Others (Rytter 1999; Vanninen and Mäkelä 1999; Mäkelä and Vanninen 2000; Al Afas et al. 2008) were unable to find any relationship between fine root biomass and easily measured tree variables and so were unable to develop a biomass function for their cases.

Chapter 8
Stand Measurement

Abstract Measurement of forest characteristics across large forest areas involves scaling up measurements from individual trees to patches of forest (stands) to large forest areas. This chapter deals with measurement of stands. Characteristics discussed are tree age, stand basal area (tree cross-sectional area at breast height summed over all the trees in a stand and expressed per unit ground area), stocking density (the number of tree stems per unit ground area), quadratic mean diameter (diameter of the tree of average basal area in the stand), site productive capacity, stand stem volume, stand biomass and stand growth.

8.1   Stands and Why They Are Measured

Until now this book has been concerned with measurement of individual trees. However, forest owners and managers need to know how much timber or other forest products are available in total from their forest estate. This allows them to determine the overall value of the estate or to work out how much wood they can cut from it year by year and still be sure that the forest will go on producing timber forever.

One way to determine the total amount of wood, biomass or other tree products in an entire forest is to measure every single tree in it and add up the results. This would be an impossibly large task for any but the tiniest patch of forest. Instead, methods are used to scale up measurements made of some individual trees in the forest to estimate what is available from the whole forest. The concept of scaling up was introduced in Sect. 1.3.

The process of scaling up involves two steps. Firstly, measurements are made of individual trees in stands. 'Stand' is a peculiarly forestry term that refers to a more or less homogeneous group of trees in a forest in which an observer might stand and look about him or her. Measurements made of stands are recorded usually on a per unit ground area basis, for example, the volume of stem wood per hectare in the stand. If the stand is on sloping ground, the area is taken to be the equivalent horizontal ground area. Stand measurements are sometimes recorded also as an average for the stand, for example, the average of the tree stem diameters.

© Springer International Publishing Switzerland 2015
P.W. West, *Tree and Forest Measurement*, 3rd edition,
DOI 10.1007/978-3-319-14708-6_8

The second step in scaling up involves measuring many stands scattered throughout the forest. The information from those many stands is then used to estimate the total amount of whatever is being measured across the entire forest. That second step is known as forest **inventory**. It will be considered in detail in Chaps. 9–11.

8.2   Measurements Taken in Stands

The definition of a stand as a more or less homogeneous patch of forest is rather loose in that it does not specify any particular area for the stand. In fact, the area will vary greatly depending on how the person measuring the forest chooses to define 'homogeneous' for the forest concerned; this will be a matter of judgement and will reflect the nature of the measurements being made and their purpose.

However the stand is defined, it will not usually be the whole stand that is measured. Rather, a sample will be selected from it for measurement; that is, a relatively small part of it will be measured and the result taken as being representative of the whole stand. Often, that part will be a plot. Commonly in forestry, such plots have an area of 0.01–0.1 ha (100–1,000 m^2), the actual area being chosen to suit the forest circumstances and the purpose of the measurement. However, as will be discussed below, there is at least one other way to measure a stand that does not involve the establishment of a plot.

The measurements made most commonly in stands are:

- Tree age
- **Stand basal area** (tree cross-sectional area at breast height summed over all the trees in a stand and expressed per unit ground area)
- Stocking density (the number of tree stems per unit ground area)
- **Quadratic mean diameter** (diameter of the tree of average **basal area** in the stand)
- Site productive capacity (defined in Sect. 8.7)
- Volume (the volume of the stem, over or under bark, or of some log product expressed per unit ground area)
- Biomass (the biomass of some part of the tree expressed per unit ground area)
- Growth (the change with time in variables such as stand basal area, stand volume or stand biomass)

Many other characteristics of stands can be, and often are, measured. However, those listed above are variables used commonly in forestry and forest science. Each will be discussed in this chapter.

8.3 Age

The age of the trees in a stand is usually most relevant when the stand is **even-aged**, that is, all the trees in the stand regenerated naturally (in native forest) or were planted (in a plantation) at or about the same time.

Even when a stand in native forest is classed as even-aged, there will have been some period of time (months to a year perhaps) over which the trees regenerated following a disturbance of the pre-existing forest that resulted in all its trees being removed or dying. That is to say, not all the trees in an even-aged native forest will be exactly the same age. For most forestry purposes, these small age differences between individual trees are ignored, and the age of the native forest stand is considered to date from the time when most regeneration took place. This is often known from historical records. Other techniques may be necessary to determine stand age when the forest history is unknown, such as counting tree rings of individual trees.

In the case of plantations, stand age is usually known from planting records. Tree seedlings are often 6–12 months old, sometimes older, when they are obtained from a nursery and planted out (West 2014a, Sect. 5.2.3). In most countries, plantations are aged from their date of planting. Scientifically, this convention seems to provide adequate results for practical purposes even though it is known that the trees are generally older than the age assigned to the plantation. In some European countries, plantations are aged from the date of sowing seeds in the nursery. In colder regions, seedlings may be raised for up to 2 years in a nursery, so prescribing a plantation age by the sowing date may then be rather more realistic biologically than using the planting date.

Many native forests are **uneven-aged**. That is, trees of a wide range of ages occur in the forest. This is usual for forests such as rainforests, where a long process of succession occurs during the life of the forest. Following a disturbance in such forest (due to things such as fire, storm or logging), certain species that respond to the full sunlight conditions created by the disturbance regenerate and grow vigorously. Other species that are tolerant of shade then regenerate and grow slowly below the canopy of the light-tolerant species. Eventually these slower-growing species reach the full sun and ultimately dominate the forest. Such forests contain trees of a wide range of ages and it is impossible to assign any particular age to a stand. However, it is often useful to know the time since the initial disturbance; this may be a useful guide to what stage of development the stand has reached. Lisa and Faber-Langendoen (2007) developed a method to define the stage of development of uneven-aged forest by measurement of various characteristics of the trees within the stand.

8.4   Basal Area

Stand basal area (stem cross-sectional area at breast height summed over all the trees in a stand and expressed per unit ground area) is important to forestry because, just as tree stem diameter is often well correlated with individual tree stem volume (Sect. 6.2.1) or biomass, stand basal area is often well correlated with stand stem volume or stand biomass.

The stand basal area of forests tends to increase with age as the trees grow. It varies also with the number of trees in the stand. For young stands, or stands with very low stocking density, stand basal area can be close to zero. For old stands and stands with high stocking densities, it can exceed 100 m^2/ha.

There are two main methods used to measure stand basal area, plot measurement and **point sampling**. These are discussed in the next two subsections.

8.4.1   Plot Measurement

Stand basal area may be determined by positioning a plot of known area somewhere within the stand. The diameters at breast height (over or under bark as required) of all the trees in the plot are then measured. The diameters are then converted to stem cross-sectional areas, the results summed and divided by the plot area to give stand basal area.

There is a general discussion of plot establishment and measurement techniques in Sect. 11.5.

8.4.2   Point Sampling

A second method of measuring stand basal area is very important in forestry, especially in forest inventory (Chap. 11). It was invented in 1948 by an Austrian forester, Professor Walter Bitterlich. In this book, the method will be termed point sampling. However, it is known also as variable radius plot sampling, variable plot sampling, angle gauge sampling, angle count sampling, Bitterlich sampling, Relaskop sampling, prism sampling or plotless sampling.

Point sampling is an extraordinarily simple method of stand measurement that does not require the establishment of a plot. It uses an instrument that can be as simple as a small stick. The measurement is made by standing at a point somewhere within the stand and simply counting the number of certain trees around the point.

The principle of point sampling is described in Fig. 8.1. This represents a view looking vertically down on a stand. Suppose an observer standing at point O is holding horizontally a small, straight stick AB at arm's length. In the near vicinity of the observer are several trees (three in the diagram, but the number is not

Fig. 8.1 Principle of point sampling

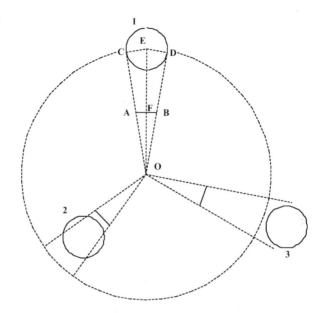

important to the argument), each with exactly the same diameter at breast height over bark; the solid circles numbered 1, 2 and 3 represent the tree stem cross sections at breast height. The centre of the stem cross section of Tree 1 is positioned at E.

Tree 1 has been positioned deliberately in the figure so that when the observer looks past the ends of the stick to breast height on that tree, the ends are aligned exactly with the width of the stem as the observer sees it; that is, lines OAC and OBD are both tangents to the circular cross section of Tree 1, touching its circumference at C and D.

The views the observer would have of Trees 2 and 3, when he or she turns in their respective directions, are illustrated also on the figure. Tree 2 is closer to the observer than Tree 1, so the stick is not wide enough to cover fully the cross section of the tree as the observer views it; the observer will see the sides of the stem projecting past the ends of the stick. Tree 3 is further from the observer than Tree 1, so the width of the stick will more than cover the width of the stem as the observer views it.

Now, suppose all the trees in the vicinity of the observer have the same diameter as the three in the diagram. Suppose the observer turns through a full circle at O and counts the number of trees he or she sees for which the width of the stick does not cover the width of the tree stem (as for Tree 2) or just exactly covers it (as for Tree 1). He or she does not count any tree for which the width of the stick more than covers the width of the tree stem (as for Tree 3). This means that any tree of the same diameter as the three trees in the figure and that is at the same distance or closer to the observer than Tree 1 would be included in the count.

Suppose the radius at breast height over bark (the radius is half the diameter) of the trees in the figure is r (cm). The radius of Tree 1 is represented by either of the lines EC or ED (geometrically, those radii will be perpendicular to the tangents to the circle OBD and OAC, respectively). Imagine that as the observer turns around at O, he or she sweeps out a circular area, shown by the large dashed circle in the figure that is of radius R (m) (the length of the line OE), and counts trees as described above. Suppose that during the sweep the observer counts n_r trees of radius r. Since all those trees are at the same distance or closer than Tree 1, their stem centres all lie within that circle of radius R.

By definition, the stand basal area of those n_r counted trees, G_r, (m²/ha) is the sum of their cross-sectional areas at breast height divided by the ground area they occupy. Knowing that the area of a circle is the product of π and the square of its radius, it follows from this definition that

$$G_r = \left[(\pi/10,000)n_r r^2\right]/\left[(\pi/10,000)R^2\right] = n_r r^2/R^2 = n_r \beta \qquad (8.1)$$

where $\beta = r^2/R^2$. The values 10,000 in this equation simply serve to convert the units of measurement (stem radii in centimetres are converted to metres and the dashed circle area in square metres is converted to hectares, there being 10,000 m² in 1 ha).

Consider now the right-angled triangle EDO (angle EDO is the right angle), where ED $= r/100$ m and EO $= R$ m. Let the angle EOD $= \theta$ degrees, so that

$$\sin(\theta) = \text{ED/EO} = (r/100)/R. \qquad (8.2)$$

Hence, given the definition of β above,

$$\beta = r^2/R^2 = 10,000 \, \sin^2(\theta). \qquad (8.3)$$

Now, suppose the stick being held by the observer is of length AB $= l_s$ (cm) and is being held a distance OF $= l_e$ (cm) away from the eye. Consider the right-angled triangle OFB (angle OFB is the right angle), where FB $= l_s/2$ and FO $= l_e$. Using Pythagoras' theorem (Appendix 4), BO² $=$ FO² $+$ FB², hence, BO $= \sqrt{[(l_s/2)^2 + l_e^2]}$. Then, since angles FOB and EOD are the same angle, θ,

$$\sin(\theta) = \text{FB/BO} = (l_s/2)/\sqrt{\left[(l_s/2)^2 + l_e^2\right]}. \qquad (8.4)$$

Hence, using (8.3) and (8.4),

$$\beta = 10,000(l_s/2)^2/\left[(l_s/2)^2 + l_e^2\right]. \qquad (8.5)$$

The stand basal area of the trees of radius r can now be determined using (8.1), that is, simply as the product of the number of trees of radius r that were counted

(n_r) and β. The importance of (8.5) is that we see that β can be determined knowing only the length of the stick the observer is holding (l_s) and the distance it is being held from the eye (l_e). That is, stand basal area can be determined with no knowledge of the tree stem radius (r) or of the radius of the large dashed circle (R) within which trees of that radius are counted.

Even more importantly, this theory works for trees of any radius, not just r. Trees with a radius larger than r will be counted within a circle of radius somewhat larger than R, whilst trees with a radius smaller than r will be counted within a circle of radius somewhat less than R. But the observer need not know the tree radius or the circle radius to apply the theory; he or she can determine β from the length of the stick and its distance from the eye. Thus, suppose the observer counts a total of n trees from the spot, a total that includes trees of any size, then the total stand basal area G (m^2/ha) will be given simply by

$$G = n\beta. \qquad (8.6)$$

Equation (8.6) is perhaps the most astonishing result that has ever been devised for forest measurement. It means that stand basal area can be measured by counting trees from a single point using an instrument as simple as a straight stick. All the observer then needs to do is measure the length of the stick and the distance it is held from the eye and then use (8.5) to determine β. In point sampling, β is known as the basal area factor; in the measurement units used here, the units of β are m^2/ha.

8.4.3 Practicalities of Point Sampling

One limitation of point sampling is that it can be used only to measure stand basal area over bark. If an under bark basal area is required, a method is needed to convert the over bark measurement to an under bark measurement. One system to do so involves measuring the over and under bark diameter of each tree as it is counted. The method discussed below (Sect. 8.8.2) can then be used to find the stand basal area under bark by replacing stem volume with stem basal area under bark in (8.10).

A number of issues need to be considered if point sampling is to be used appropriately. A straight stick is a perfectly good 'instrument' to use for point sampling, as indeed is the width of the observer's thumb. Thumbs generally have a basal area factor within the range 2–5 m^2/ha depending on the individual's thumb width and the length of his or her arm; using (8.5), most foresters have worked out the factor for their thumb and use it to assess, quickly and easily, the stand basal area of a piece of forest they are visiting.

Often, a small triangularly shaped glass or plastic prism (or basal area wedge as they are often called) is used to carry out point sampling. When a tree stem is viewed through the prism, the triangular shape of the prism causes the view of the stem section to be displaced sideways due to light refraction by the material of the prism. If, as the observer views it, the stem section is displaced less than, or just as much as, the width of the stem, the tree is counted. Otherwise, it is not. The principle of the method remains exactly the same, but the angle of the triangle of the prism and the refractive qualities of the material from which it is made together determine the basal area factor of the prism. Forestry suppliers sell these prisms with a wide range of basal area factors. The main advantage of prisms is that it is easier to see the tree stems than it is with a stick or thumb and so easier to judge whether a particular stem should be included or not in the count.

The instruments used for optical dendrometry (Sect. 5.3.4) are usually constructed to allow point sampling to be carried out with them. Various other (often very cheap) devices for point sampling are available also from forestry suppliers.

There are always trees for which the observer will find it difficult to judge whether they should or should not be included in the count. These will be trees that are near the circumference of a circle being implicitly swept out by the observer, such as Tree 1 in Fig. 8.1. When such cases arise, the diameter of the tree at breast height over bark and the distance from its stem centre to the observer should be actually measured. If its diameter is D (cm) and the distance to it is d (m), then it follows from the mathematics of Sect. 8.4.2 that the tree should be counted if $d \leq D/(2\sqrt{\beta})$ and otherwise not (this computation determines the radius, R, of the circle being swept out implicitly by the observer for trees of that particular diameter).

If the ground on which the count is being made is sloping, some of the instruments available for point sampling correct automatically for the slope. If using an unsophisticated instrument like a stick, it will be necessary to measure the diameter of each tree counted (D, cm) (or at least of those trees that are doubtful), the distance along the slope from the observer to the tree (s, m) and the slope angle (γ, degrees). The corresponding horizontal distance to the tree (d, m) is then calculated as $d = s \cos(\gamma)$. The expression given in the previous paragraph can then be used to determine if the tree should be counted or not. Generally, this is necessary only if the slope exceeds about 8–9°.

It is important too that an appropriate basal area factor is used for the stand being considered. If the factor is too small, then a very large number of trees will be counted and some of the ease and speed of the method will be lost. If the factor is too large, too few trees will be counted to be a reasonable sample from the stand and the basal area determined may not represent the stand adequately. The larger the trees in the stand, the larger will be the appropriate basal area factor. In general, it might usually be considered appropriate to use a factor size that leads to a count of around ten trees. Many of the instruments available to carry out point sampling incorporate several factors, and the user may choose one that is appropriate to the stand being considered.

Particular care needs to be taken in densely stocked forest where some trees may be obscured from the observer's view by others. Also, it is easy for the observer to lose track of which trees have been considered and which have not. Under these circumstances, it is desirable to have a two-person team to carry out the measurements. One acts as the observer and the other walks about the stand keeping track of what the observer has and has not viewed. To view obscured trees, the observer may have to move slightly away from the sampling point, but must ensure that he or she remains the same distance away from the obscured tree.

8.4.4 Plot Measurement Versus Point Sampling

It seems amazing that stand basal area can be measured with point sampling by merely counting trees from a single point using an instrument as simple as a small stick. Even after they have tried it, my forest measurement students always have great difficulty believing that point sampling really works even if they have studied the theory in Sect. 8.4.2; it just seems too simple to be true. On the other hand, they have no difficulty accepting the results from measuring a plot (Sect. 8.4.1); they can see the plot boundaries and the trees in it that they have measured with a diameter tape. Somehow, point sampling seems to be cheating; trees are just counted, not measured physically.

For both plot measurement and point sampling, one issue of concern is trees that are close to the boundary. In plot measurement, the convention adopted usually is that a tree is considered to be in the plot only if the centre of its stem lies within the plot boundary. The method used to decide if a tree should or should not be counted in point sampling was discussed in Sect. 8.4.3; similarly to plot measurement, this method includes trees for which the position of the centre of their stem cross section lies within the implicit circular area being swept out in taking a point sample (Fig. 8.1).

It is important to realise that point sampling and plot measurement will usually give somewhat different answers for stand basal area; the answers will usually not be vastly different, but certainly often differ by several per cent. In point sampling, the size of the circle effectively swept out by the observer in making the count differs for the different diameters of the trees being counted (circle sizes that the measurer need never know) (Fig. 8.1). Hence, it is impossible in this case to define exactly what area of the stand is being used to measure the stand basal area; this has led to one of the alternative names used for point sampling, variable radius plot sampling. On the other hand, with plot measurement, a particular section of the stand is clearly identified as the part being measured.

As well, it must be appreciated that both point sampling and plot measurement are a form of sampling of the stand (Sect. 8.2). An arbitrary choice has to be made as to the location of the plot in the stand or of the point at which the point sample is taken. In other words, plot measurement and point sampling will take samples from the stand in rather different ways. Inevitably then, they will give somewhat different results for stand basal area. That is not to say either result is wrong or one of the results is generally better, or more useful, than the other. They are just different results that reflect the fact that the two methods sample the stand in different ways.

8.5   Stocking Density

Both plot measurement and point sampling can be used to determine stand stocking density (the number of tree stems per unit area). In plot measurement, the number of trees in the plot is counted. The number divided by the area of the plot is then the stocking density.

To get the stocking density when a point sample is taken, it is necessary to not only count the trees to be included in the point sample but also to measure the diameter at breast height over bark of each counted tree. If there were n trees counted and the diameter of the ith of those trees was D_i (cm), then the stand stocking density (S, stems/ha) can be determined as

$$S = \beta(40,000/\pi) \left[\Sigma_{i=1...n} \left(1/D_i^2 \right) \right] \tag{8.7}$$

where β is the basal area factor (m^2/ha) used in the point sample. Note that in this equation, the mathematical expression $\Sigma_{i=1...n}$ denotes the summation of the term that follows the expression, as i takes successive values 1, 2, 3... etc., up to n.

There are methods other than point sampling that also do not require the establishment of a plot and that can be used to estimate stocking density. These involve selecting a number of points at random within the stand and measuring the distance from them to neighbouring trees. Payendah and Ek (1986) describe these methods. These techniques have been subject to bias and rather more uncertainty than counting trees in a plot or doing point sampling; however, research has being addressing these limitations (Picard et al. 2005; Kleinn and Vilčko 2006a, b; Magnusson et al. 2008).

Stocking density is useful for three main purposes. Firstly, dividing some other stand estimate (say, stem volume per unit area) by stocking density gives the average volume per tree in the stand. Secondly, the stocking density of a stand is an important variable used in describing the stage of development of a stand; stocking density changes with age during the life of a stand as trees die or new seedlings are recruited to the stand. Determining how and when changes in stand stocking density occur is an important part of forest growth modelling, where mathematical models are used to predict how stands will grow and change with time. The use of growth models is fundamental to modern forest management and allows managers to predict the availability of wood, or other forest products, from a forest far into the future. Growth modelling will not be discussed further in this book; there are various reviews and examples of both older and more recent approaches to forest growth modelling (Vanclay 1995; Battaglia and Sands 1998; Mäkelä et al. 2000; Peng 2000a, b; Le Roux et al. 2001; Avery and Burkhart 2002; Porté and Bartelink 2002; Pretzsch et al. 2002; Landsberg et al. 2003; Valentine and Mäkelä 2005; Cienciala and Tatarinov 2006; Richardson et al. 2006; Tatarinov and Cienciala 2006; Fourcaud et al. 2008; Weiskittel et al. 2011).

8.6   Quadratic Mean Diameter

The average of the diameters of the trees in a stand is often a quite useful measurement to characterise the condition of a stand. Another measure used is called the quadratic mean diameter. This is the diameter corresponding to the average basal area of the trees in the stand. If a stand has a basal area G (m^2/ha) and a stocking density of S stems/ha, then its quadratic mean diameter D_q (cm) is

$$D_q = \sqrt{[(40,000/\pi)\, G/S]}. \qquad (8.8)$$

It is argued that quadratic mean diameter is often more useful than average diameter, because it relates more closely to stand volume. It is also a useful measure to give some idea of the size of the trees in a stand when their average diameter has not been recorded. It has other uses, especially in defining the density of stands (the degree of crowding of the trees) (e.g. Pretzsch and Biber 2005; Woodall et al. 2005; VanderSchaaf and Burkhart 2007). Measures of stand density will not be discussed further here; their theoretical and biological bases are discussed in West (1983, 2014a, Sect. 7.1) and in other forest measurement texts (e.g. Avery and Burkhart 2002).

8.7   Site Productive Capacity

In all editions of this book (since 2004), I have defined 'site productive capacity' as the total stand biomass produced by a stand on a particular site, up to any particular stage of its development, when the stand has been using fully the resources necessary for tree growth that are available from the site. More recently Skovsgaard and Vanclay (2008) defined it 'as the potential of a site to produce plant biomass'. Whilst these are formal definitions, we can think of site productive capacity more simply as how fast trees can grow on a site.

Stand biomass production at any time is limited by the genetic characteristics of the species concerned, by the environmental characteristics of the site on which the stand is growing (particularly the climate and soil fertility), by the stage of development of the stand, by the stocking density of the stand and by the way in which the trees have been tended (the **silviculture** applied) (Skovsgaard and Vanclay 2008; West 2014a).

Measurement of site productive capacity is very important to forestry because it indicates the maximum amount of biomass or of the principal forest product, wood, that might be available from a forest on the particular site on which it is growing. Reviews of the concepts surrounding measurement of site productive capacity can be found in Skovsgaard and Vanclay (2008) and Bontemps and Bouriaud (2014).

8.7.1   Geocentric and Phytocentric Measures

Many attempts have been made to measure the characteristics of the soil and climate at a site and use this information to predict forest production. Some work of this type has related observed growth of forest stands to observed environmental characteristics, both weather and soil (Turvey et al. 1990; Osler et al. 1996b; Hackett and Vanclay 1998; Snowdon 2001; Uzoh 2001; Ryan et al. 2002; McKenney and Pedlar 2003; Kimsey et al. 2008; Latta et al. 2009; Sabatia and Burkhart 2014; Saremi et al. 2014a, b). Other work has involved the development of complex models that predict the physiological response and growth of tree stands in relation to environmental characteristics (Running 1994; Coops et al. 1998; Sands et al. 2000; Ditzer et al. 2000; Mäkelä et al. 2000; Mummery and Battaglia 2001; Landsberg et al. 2003; Roxburgh et al. 2004; Buckley and Roberts 2005; Swenson et al. 2005; Cienciala and Tatarinov 2006; Louw and Scholes 2006; Tatarinov and Cienciala 2006; Waring et al. 2014).

Because they are based on measurement of site environmental characteristics, these methods of measuring site productive capacity have been termed 'geocentric' methods. They have been used principally to develop systems to assess whether or not particular sites are appropriate for afforestation. The methods are sufficiently complex to warrant books of their own (e.g. Landsberg and Sands 2011; Weiskittel et al. 2011) and will not be considered further here.

An alternative to geocentric measures of site productive capacity are what are known as 'phytocentric' measures. These are based on measurement of the growth of the trees that has actually occurred on a site, rather than on measurement of the environmental characteristics of the site itself. Obviously then, a phytocentric method can be used only after the forest has been established on the site and only after sufficient time has elapsed for sufficient growth to have occurred to provide a suitable measure of growth on that site. Thus, a phytocentric measure might be obtained relatively early in the life of a forest and then used to predict how well the forest will grow to later ages.

Because they are based on the actual growth of the forest on a particular site, it can be expected that phytocentric site productive capacity measures should be unbiased and very precise estimators of forest growth. However, they are unable to predict how well forest will grow on a site before it is growing there; only geocentric measures or geocentric methods allow prediction of growth before the forest is established.

A particular phytocentric measure of site productive capacity has been used in forestry for many years in the management of even-aged forests. The measure is known as **site index**. It is based on measurement of the taller trees in a stand to determine what is known as stand **dominant height**. Section 8.7.2 will describe how and why this height measurement is made, whilst Sect. 8.7.3 will consider site index itself.

Forestry science has paid much less attention to the development of measures of site productive capacity for uneven-aged forests. The inability to define their age makes it impractical to use the types of measure, such as site index, that have been developed for even-aged forest. One phytocentric measure that has been used for uneven-aged forests is the stand basal area or dominant height when the forest has reached an equilibrium stage of development (see Sect. 8.10) after which its basal area or height does not change with time. Other measures include the height of trees with a specified stem diameter or a complex summation of the diameters of specific 'index' species in the stand. These measures apply generally to the later developmental stages of uneven-aged forest (Sect. 8.10) and it would be difficult to use them to measure site productive capacity when the forest is at an early stage of its successional development. Vanclay (1992) has reviewed these measures and work continues on their development and with other approaches (Pokharel and Froese 2009; Berrill and O'Hara 2014). Work to develop geocentric measures for uneven-aged forest (e.g. Ditzer et al. 2000; Weiskittel et al. 2011) may provide eventually better measures of site productive capacity than the phytocentric measures that have been developed to date.

8.7.2   Dominant Height

The average height of the trees in a stand can be a useful measure of stand condition. More frequently in forestry, measures known by the generic term stand dominant height are used to represent stand height. These are defined generally as the average height of the tallest trees in the stand.

In an even-aged forest, the sizes of individual trees differ because they compete with each other for the resources of the site they require for growth, that is, light, water and soil nutrients. The more successful competitors become larger in size by suppressing the smaller less competitive trees. The intensity of this competition depends both on the degree of crowding of the trees on the site and the rate at which the more successful competitors grow. The more the crowding and the faster the more successful trees grow, the more rapidly will the smaller trees succumb to the competitive pressure and eventually die.

Given this, it is the characteristics of the more successful trees that best reflect the site productive capacity. Because the larger trees suppress the smaller ones, the characteristics of the smaller trees merely reflect the fact that they are the less successful competitors. Furthermore, it is height, rather than biomass, that is the characteristic of the competitively more successful trees that best reflects site productive capacity. Their biomass will depend very much on the stocking density of the stand in which they are growing. If, for whatever reason, the stocking density happens to be low, they may not be using fully the resources for growth available from the site; their biomass would then be lower than it would be if there were more of them in the stand. However, much research has shown that their heights are little affected by their stocking density (MacFarlane et al. 2000; Henskens et al. 2001; Antón-Fernández et al. 2011; Xue et al. 2011); their heights will reflect the site productive capacity, even when the stocking density is sufficiently low that stand biomass is not reflecting it. These arguments show why it is the height of the taller trees in stands, that is, stand dominant height, that relates best to the productive capacity of a site and so has become important as a measure of site productive capacity.

There are various ways in which stand dominant height is defined. One measure, **top height**, is defined as the average height of a specified number per unit area of the trees of a stand with the largest diameters at breast height (usually over bark). A second, **predominant height**, is defined as the average height of a specified number per unit area of the tallest trees in the stand. Other names and slightly different definitions are often used for these measures, but the terms and definitions used here are perhaps the most common.

It is up to the measurer as to the number per unit area of largest diameter or tallest trees to be used in these definitions for a particular forest type. The numbers used in practice vary widely from country to country and forestry organisation to organisation. Usually, a value within the range 40–100 stems/ha is chosen. The different numbers used in different places often make it difficult to compare published results from different parts of the world.

Top or predominant height is measured in a plot of known area within a stand. Say the plot was 0.05 ha in area (perhaps a rectangular plot of 20×25 m) and the definition of top or predominant height to be used specified 40 stems/ha to be measured. Then, the heights of the $40 \times 0.05 = 2$ largest diameter or tallest trees would be measured in that plot and their average used as the measure of top or predominant height, respectively. If the plot size and number per hectare are such that a non-integer number of trees should be measured, then the number actually measured is usually taken as the nearest integer number.

Practically speaking, it is easier generally to determine top height than predominant height because it is easier to identify the largest diameter trees in the stand than the tallest ones; diameters of all the trees in a plot are often measured anyway. Of course, the largest diameter trees in the stand are often the tallest in any case; in practice, the difference between top height and predominant height is usually small for most forest stands.

8.7.3   Site Index

The dominant height of a forest stand changes with age as it grows. To allow for this, the phytocentric measure of site productive capacity used in forestry, site index, is defined as stand dominant height at a particular age. For any particular type of forest in any particular forest region, the age chosen to define site index is not of any special importance; it is chosen entirely at the discretion of whoever develops the measure for that forest. Whatever age is chosen, it is termed the index age for that forest. It is also up to the developer of the index to decide if top or predominant height is to be used as the measure of dominant height.

Thus, suppose the index age chosen for a particular forest type is 20 years. Then, a stand with a dominant height of 25 m at 20 years of age would have a site index of 25 m. For even-aged forests throughout the world, research has shown consistently that site index is a very reliable measure of site productive capacity. The only *caveat* on this is that it is true only after the forest reaches about 5–10 years of age. In forest younger than this, the trees may not have grown large enough to be competing with each other, so that dominant height becomes a good measure of productive capacity (Sect. 8.7.2).

One way to measure site index is to wait for a stand to reach the index age, and then measure its dominant height. However, for most of the commercially important forest types throughout the world, site index functions have been developed so that stand site index can be determined for the stand, no matter at what age it is measured. A typical example will be described here to illustrate how site index functions are used.

The example is taken from West and Mattay (1993), who developed a site index function for flooded gum (*Eucalyptus grandis*), a species that grows in even-aged native and plantation forests in subtropical eastern Australia. West and Mattay defined stand top height in that forest as the average height of the 50 largest diameter trees/hectare and site index as the top height at 20 years of age (the index age they chose). Their function allows prediction of the top height H_2 (m) of a stand at some age A_2 (years) from its top height H_1 (m) that was measured at some other age A_1 (years), as

$$H_2 = H_1 \left\{ [1 - \exp(-0.0126A_2)] / [1 - \exp(-0.0126A_1)] \right\}^{0.563}. \qquad (8.9)$$

Suppose a flooded gum stand was measured at 12 years of age and found to have a top height of 23.1 m. Equation (8.9) predicts that its top height would be 30.0 m at 20 years of age. Since 20 years of age was the index age used for these forests, the site index of this stand has then been estimated as 30.0 m from a measurement of its top height at 12 years of age. The function could be used in a similar fashion to predict the stand site index from measurement of its top height at any other age.

Figure 8.2 shows how West and Mattay's site index function predicts stand top height will change with age in stands of site index 20, 30 or 40 m. Note that at the index age chosen for this forest, 20 years of age, the function predicts (by definition) that top height equals the site index. Similar lines could be drawn for any other site index. The position is indicated on the 30 m site index curve of the top height H_2, as predicted from the measured top height H_1, in the example in the preceding paragraph.

Many functional forms different from that of (8.9) have been used as site index functions by different authors from time to time for various species in various parts of the world. Huang (1997) listed a number of these alternatives. However, (8.9) is a commonly used function that continues to be found useful (Diéguez-Aranda et al. 2006a; Louw and Scholes 2006; Mamo and Sterba 2006; Nord-Larsen 2006).

Fig. 8.2 Change with age of top height of even-aged stands of *Eucalyptus grandis* of site index (SI) 20, 30 or 40 m predicted using the site index function of West and Mattay (1993). The *vertical dashed line* shows the index age they used to define site index in this forest. In the example discussed in the text, the value of top height H_2 was predicted from a measured value H_1 using (8.9)

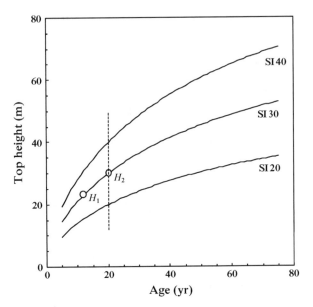

The importance that the use of site index has assumed in even-aged forest management cannot be overemphasised. It is an integral and basic part of most of the forest growth models that are used by managers of even-aged forests throughout the world to predict the long-term availability of wood from them. Skovsgaard and Vanclay (2013) have discussed some issues surrounding the use of site index and considered some new approaches to its use.

8.8   Volume

The way in which stand stem volume is measured will depend on the resources that are available to the measurer and the precision required of the estimate. This section will consider first the case that stand stem volume is being measured in a plot and then the case that the stand is being measured using a point sample.

8.8.1   Plot Measurement

A first possible way to determine stand volume for a plot, and by far the most laborious, would be to measure directly the stem volume of each and every tree in the plot. The individual tree volumes would then be added up and the result divided by the plot area to give stand volume. The tree volumes could be found using the methods for sectional measurement of standing trees (Sect. 5.3.4) or through centroid sampling (Sect. 5.4). Stand volumes, either as total stem volume or volumes of different merchantable log sizes, could be obtained with these methods. Allowance must be made if under bark, rather than over bark, stem volumes are desired; this issue was discussed in Sects. 5.3.4 and 5.4.

A second method would be to measure directly the volumes of only a sample of trees from the plot. The information from that sample could then be used to estimate the stand stem wood volume for the whole plot. There are a number of mathematically formal ways in which the trees to be included in the sample could be chosen. These are discussed in detail in Chaps. 9 and 10; examples are given in Sects. 9.4, 10.3 and 10.4 that describe the estimation of stand stem wood volume from a sample of trees taken from a plot.

A third method would also involve measuring volumes of a sample of trees from the plot. For many stands, a graph of stem volume against stem diameter at breast height will show there is a very close relationship between these two variables for the sample trees. A regression equation (Sect. 6.2.1) could then be fitted to the sample data and this used to predict volumes of all other trees in the stand from measurement only of their stem diameters. This method is known as model-based sampling; an example where it is used to determine stand volume is given in Sect. 10.5.

If a tree volume or taper function (Chap. 6) is available for the species concerned, a fourth method to determine stand volume would be to use those functions to estimate total or merchantable stem volumes for each and every tree in the plot. This usually requires that only tree diameter at breast height and total height of each tree be measured (or occasionally some additional measurements, as in the examples in Sect. 6.2.1).

Measurement of the height of every tree in the plot, as required by the fourth method, can be time consuming. Particularly for even-aged forests, there is often a close relationship between tree diameter at breast height and total height. Where this is the case, a fifth method to estimate stand volume might be used. In this case, a sample of trees from the plot would be measured for height. Then, a regression relationship could be established from the sample data, from which other tree heights in the plot might be estimated from their measured diameters. West (1979) discussed a number of regression functions that might be suitable for this purpose. Soares and Tomé (2002) also listed possible functions to do this and considered also functions that might be used to predict individual tree heights generally, for a particular species right across a particular region.

It would be expected that the precision of the estimates of stand stem volume made with these various methods would decline in much the same order as they have been presented here. More and more assumptions and regression functions are involved the further one moves through the five methods. Interestingly however, West (1979) made a study of the precision of the estimates of stand total stem wood volume obtained in stands of 20–100-year-old native eucalypt forests in Tasmania, Australia, using what might be expected to be the least precise of all these methods, the fifth. He found that even for this method, the stand volume estimated nearly always lay within ±6 % of the true stand volume and was usually much closer.

This result illustrates how reliable are the methods that have been devised for measuring stand wood volumes in forests.

A sixth possible method for measuring stand volume is worth mentioning also. Stand volume estimation functions have been developed from time to time for particular forest types in particular regions. Rather than having to measure individual trees in a stand, these functions allow estimation of stand volume (total stem or merchantable volume) directly, usually from measurement of stand basal area and stand dominant height. Whilst becoming less common today, stand volume estimation functions do continue to be developed (Brooks and Wiant 2004; Vincent et al. 2014).

8.8.2   Point Sampling

If a point sample is being taken in a stand (Sect. 8.4.2), its method can be adapted to provide an estimate of stand volume. To do so requires measuring the diameter at breast height over bark and stem volume (total or merchantable) of each of the trees counted in the point sample. The volume measurements could be done by direct measurement with sectional measurement (Sect. 5.3), by centroid sampling (Sect. 5.4) or by estimation with a tree volume or taper function (Sects. 6.2 and 6.3).

Suppose n trees were counted in the sweep, the diameter of the ith of those trees was measured as D_i (cm), and its stem volume was measured as V_i (m^3). Then, the stand stem volume V (m^3/ha) can be determined as

$$V = \beta(40,000/\pi) \left[\Sigma_{i=1...n} \left(V_i/D_i^2 \right) \right] \qquad (8.10)$$

where β (m^2/ha) is the basal area factor. Marshall et al. (2004) have considered what basal area factor is appropriate to use in such cases.

8.9   Biomass

The biomass of the whole or parts (leaves, stems, roots, etc.) of individual trees in a stand may be obtained either by direct measurement (Sect. 7.2) or through the use of biomass estimation functions (Sects. 7.3 and 7.4). Because of the large amount of work involved, direct measurement will be rare; usually it will be done only in a research context to obtain data to develop biomass estimation functions.

If plot measurement is being used, once individual tree biomasses have been obtained, they may be added and the total divided by the plot area to obtain the stand biomass, just as for stand basal area (Sect. 8.4.1) or stand volume (Sect. 8.8.1). If point sampling has been used, the same method is used to determine stand biomass as described for stand volume (Sect. 8.8.2), with individual tree biomasses replacing individual tree volumes in (8.10). If leaves are being considered, it is often their area, rather than their biomass, that is of interest (Sect. 7.2.1). Stand leaf area is known as **leaf area index**; it can be determined from individual tree leaf areas just as for biomass. Its value varies over a range of about 2–11 m^2/m^2 for forests of the world (Beadle 1997) and correlates closely with the availability of water or nutrients on a site. There are a number of ways by which it can be measured directly in a stand; these are forms of remote sensing and are described in Sect. 13.1.2.

An alternative to using individual tree biomass estimates is to use a function that predicts stand biomass from some more easily measured stand variable, just as has been done for stand volume (Sect. 8.8.1). Mosseler et al. (2014) gave an example for some North American willow species being grown in Canada, where stand above-ground biomass was predicted from stand average tree stem diameter or stem length. In their case, the biomass estimation functions they determined were quite specific to the site where the trees were growing. Vincent et al. (2014) developed a function for rainforests of coastal French Guiana that predicted stand above-ground biomass from stand stocking density and quadratic mean diameter. Hutchison et al. (2014) developed a function for mangrove forest scattered around the world that predicted their above-ground stand biomass from environmental variables that reflected the temperature and rainfall at the site where they were growing. Asner et al. (2012) developed a function that could be applied widely across tropical forests, based on data obtained from forests in all of Panama, Peru, Madagascar and Hawaii. It required remotely sensed stand height information obtained using airborne laser scanning (Sect. 13.2.2).

Another form of stand biomass function uses a biomass expansion factor (cf. Sect. 7.4.2). This is the ratio between stand biomass and stand volume; the aim of using such factors is to take advantage of the many tree volume functions that are available already (Chap. 6) and that can be used to give estimates of stand volume. There are a number of examples of the use of stand biomass expansion factors in various forest types (Grierson et al. 1992; Lehtonen et al. 2004, 2007; Van Camp et al. 2004; Jalkanen et al. 2005; Cienciala et al. 2008).

8.9.1 Root Biomass

As discussed in Sect. 7.2.3, there are particular problems involved with root biomass measurement, especially the difficulty of identifying to which tree in a stand any particular root belongs. When great care has been taken to match roots with trees, individual tree biomass estimation functions for roots have been developed (Sect. 7.4.4). However, to avoid the problem of having to match roots and trees, it has become common to develop stand-based biomass estimation functions for roots, in preference to functions for individual trees.

Firstly, we will consider such functions for coarse or total (coarse plus fine) stand root biomass. The most common form of these functions is the simple allometric equation (Sect. 7.3)

$$B_R = \gamma B_A{}^\delta \qquad (8.11)$$

where B_R (tonne/ha) and B_A (tonne/ha) are stand coarse or total root oven-dry biomass and stand above-ground biomass, respectively, and γ and δ are parameters of the equation. In practice, individual tree biomass estimation functions (Sect. 7.3), individual tree biomass expansion factors (Sect. 7.4.2) or stand biomass expansion factors will be used to estimate stand above-ground biomass for the trees of a plot. That estimated value of stand above-ground biomass will then be used with (8.11) to estimate a corresponding coarse or total root stand biomass.

Mokany et al. (2006) attempted to develop a widely applicable stand root biomass estimation function of the form of (8.11). They collated root and above-ground stand biomass data that had been collected by many different researchers from a wide range of forest and woodland stands of many different species and forest types, right across the world. Their function is illustrated by the solid line drawn in Fig. 8.3 that shows how the proportion of stand total biomass (above plus below ground) that is roots varies with the total biomass. Mokany et al.'s function

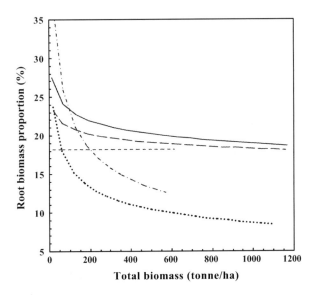

Fig. 8.3 Stand root biomass as a proportion of stand total biomass in relation to total biomass. Results are shown for an average for forests and woodlands of the world (——) (0.489, 0.890) (Mokany et al. 2006) and (— —) (0.372, 0.924) (Niklas 2005, Table 1, Canell data set), for Canadian softwood (- - - -) (0.222, 1) and hardwood (— - —) (1.576, 0.615) forests (Li et al. 2003) and for Australian forests and woodlands (.........) (0.677, 0.712) (Snowdon et al. 2000, Table 3.6). Results were derived from functions of the form of (8.11); in each case, values in parentheses are for the parameters γ and δ of that function

predicts that stand root biomass declines progressively from around 25–30 % of total biomass, in forests with a small total biomass (say, younger forest or open woodlands), to just under 20 % in forests with a large total biomass (say, tall, mature forests). As shown also in Fig. 8.3, Niklas (2005) also used a data set collected widely across the world and obtained results similar to those of Mokany et al.

Mokany et al.'s and Niklas' results represent averages for forests generally around the world. However, there will be much variation around those averages, if one considers stands of particular tree species, of particular ages and growing in the environmental circumstances of any particular geographical region. Li et al. (2003) collated above- and below-ground stand biomass data available from the literature for forests of cold temperate and boreal regions, principally of the northern hemisphere. Their root biomass estimation functions that also used (8.11) are illustrated also in Fig. 8.3. Their results are rather different from Mokany et al.'s and Niklas' results for forests of the world generally. In particular, they found a distinct difference between forests of softwood and hardwood species; neither Mokany et al. nor Niklas identified any such difference. Results obtained using (8.11) by Snowdon et al. (2000) for a wide range of forest types right across Australia are shown also in Fig. 8.3. The Australian data were probably dominated by hardwood species, and there does appear to be some commonality between Snowdon et al.'s results and Li et al.'s hardwood results for northern hemisphere forests. However, the results do suggest that roots tend to make up a rather lower proportion of total biomass in Australian forests when compared with other parts of the world.

Research work continues to develop root biomass estimation functions of this nature, for other forest types in other parts of the world, and to study what determines the proportion of stand biomass that is roots (Kajimoto et al. 2006; Zerihun et al. 2006; Cheng and Niklas 2007; Wang et al. 2008b). No consistent pattern has emerged yet to allow the development of definitive stand root biomass estimation functions.

8.9.2 Fine Root Biomass and Area

There are perhaps even greater difficulties in measuring fine roots and determining estimation functions for them than for total or coarse root biomass (Sects. 7.2.3 and 7.4.4). Just as with total and coarse roots (Sect. 8.9.1), there have been attempts to develop functions to estimate stand fine root biomass directly rather than individual tree fine root biomass.

One approach was adopted by Zerihun et al. (2007) for woodland forests of poplar box (*Eucalyptus populnea*) in northern Australia. Assumptions were made about how fine roots were distributed spatially around individual trees in a stand and a model was then developed that predicted this distribution as a function of tree diameter. Applying this model to all trees in the stand gave an estimate of stand fine root biomass. The system was only partially successful in the woodlands where it was tested.

Other workers have found **correlation**s between fine root biomass in stands and stand parameters, such as the amount of nutrients in litter falling from the above-ground parts of trees and site rainfall and temperatures (Vogt et al. 1998); these relationships tend to be specific for the site and species concerned and cannot be used to make predictions for other sites or species.

Just as for leaves (Sect. 7.4.4), the area of fine roots can be of great interest as well as their biomass. Fine roots absorb water and nutrients through their surfaces, and so their area is an important measure of their capability to do so. Stand fine root area index is the below-ground equivalent of stand leaf area index. Al Afas et al. (2008) found that it correlated quite closely with leaf area index in a set of plots of various poplar clones (*Populus* spp. and hybrids) growing at one site in Belgium. Since it is the fine roots that must gather, from the soil, the water and nutrients required by the leaves, it might indeed be expected that fine root area index and leaf area index should be correlated. In the 2-year-old stands Al Afas et al. considered, fine root area index could be estimated quite closely as being 64 % of leaf area index. It is unlikely that the same relationship would hold on different sites and would have to be evaluated separately for each site.

All these examples of stand fine root biomass (or area) estimation functions are rather specific to the site and forest circumstances for which they were derived. There remains much research to be done before satisfactory fine root biomass estimation functions become available generally.

8.10   Stand Growth

The rate of growth of trees in forests, hence, the rate at which they produce wood or biomass, is of prime concern to forest scientists or anyone growing forests for commercial purposes. The rate of production of forests depends on the site productive capacity (Sect. 8.8), their stocking density and the way in which the trees are tended. Detailed discussion of these issues, at least for plantation forests, can be found in West (2014a, Chap. 3) and they will not be considered further here. However, there are some conventions used in forestry to refer to stand growth rates. These will be described here.

Figure 8.4a shows an example of how stand stem wood volume changes with age in a forest. The example is taken from West and Mattay (1993), for a typical stand of the same flooded gum (*Eucalyptus grandis*) forest for which the example site index function was described in Sect. 8.7.3. Growth is shown only from 5 years of age, because that was the youngest age for which West and Mattay had data available. The sigmoidal (S-like) shape of the curve is common to even-aged forests throughout the world; in fact, the growth of many biological organisms displays such a shape. In the present example, the S shape is rather asymmetrical, with the

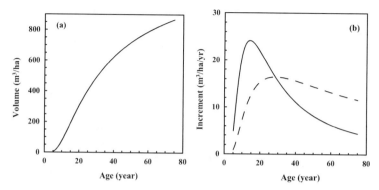

Fig. 8.4 (a) Change with age in stand stem wood volume of a typical stand in *Eucalyptus grandis* forest in subtropical eastern Australia. (b) The change with age in current annual increment (*solid line*) and mean annual increment (*dashed line*) in stand stem wood volume for the same stand (drawn using Eqs. 8, 10 and 11 of West and Mattay 1993)

bend in the S occurring at about 15 years of age. If, instead of volume, stand stem basal area or biomass was shown on the graph, the shape of the curve would be much the same.

In forestry, stand growth rate is usually expressed in one of two ways. The first, **current annual increment** (often abbreviated as CAI and also termed **periodic annual increment**, abbreviated to PAI), is the immediate growth rate of the stand at any particular age, that is, the growth rate occurring at a particular instant in time. It was this that was being referred to in using the term growth rate in the preceding paragraphs.

Trees grow too slowly to measure their growth over periods of a few seconds, as would be required in practice to determine their immediate growth rate. Hence, current annual increment is usually determined approximately by measuring a stand at intervals often of about 1 year (often longer in very slow-growing forests). Current annual increment is then approximated as the change in the stand between the two measurements, divided by the length of time between them. Figure 8.4b shows how the current annual increment of the stand in Fig. 8.4a changed with age. Note that the current annual increments shown in Fig. 8.4b are exact, because they were determined from the curve in Fig. 8.4a using the mathematical technique known as differential calculus; had they been determined by measuring the forest at annual intervals, they would have been approximate. The less mathematically inclined reader may ignore this, but should concentrate on the shape of the current annual increment curve. It shows that the current annual increment in this stand increased progressively to reach a maximum at about 15 years of age (that maximum actually occurs at the bend in the S shape in Fig. 8.4a). Thereafter, the current annual increment declined steadily. If West and Mattay had data from stands much older than the 75 years of age shown in the figure, it would be expected that their current annual increment would have declined eventually to near zero as the trees became senescent.

This pattern of growth rate is common to forests throughout the world. It would be of considerable interest to forestry to know what causes the decline in current annual increment that is apparent after 15 years of age in Fig. 8.4b. If it did not happen, the total production by forests would reach much higher levels, at much earlier ages. The principal theory to explain the decline is known as the 'hydraulic limitation' theory. This proposes, firstly, that as a tree grows taller, there is a longer path through which water must travel from the roots to the leaves; this offers an ever-increasing resistance to the flow, due to friction with the walls of the wood cells through which the water is passing. Secondly, the taller the tree, the greater is the gravitational effect resisting the upward movement of water. These effects lead to a higher level of water stress in the leaves from time to time, reducing the amount of photosynthesis and hence production that occurs in the leaves. Ryan et al. (2006) have reviewed the research that has been done to test this theory. It appears to hold well for many tree species under many circumstances, but a number of issues around it remain to be explained. Research has continued to test the theory further (Martínez-Vilalta et al. 2007; Bond et al. 2007; Mencuccini et al. 2007; Vanderklein et al. 2007; Domec et al. 2008; Nabeshima and Hiura 2008; Sperry et al. 2008; Ambrose et al. 2009; Fernández and Gyenge 2009; Drake et al. 2010, 2011; Patankar et al. 2011; Piper and Fajardo 2011; Cramer 2012; Räim et al. 2012; Xu et al. 2012).

A second measure used to describe stand growth rate in forestry is called **mean annual increment** (often abbreviated as MAI). This is the average rate of production to any particular age of the stand. It is determined simply as the stand volume (or basal area or biomass) at any age divided by the age. It is probably the most popular measure used by foresters to indicate how fast a forest grows. Mean annual increment changes with age during the life of the forest, as illustrated in Fig. 8.4b. The units for mean annual increment are m^3/ha/year, the same as those for current annual increment.

The most productive forests in the world are plantations, because they receive intensive silvicultural management. West (2014a, Sect. 3.2) has summarised knowledge on their growth rates. Plantations being grown for wood for papermaking or to be sawn into timber are usually planted at stocking densities of about 800–2,500 stems/ha. The fastest-growing hardwood plantations of these have a mean annual increment that rises to a maximum of about 60 m^3/ha/year at around 6–7 years of age and then declines steadily to just over 30 m^3/ha/year by 30 years of age. Softwood plantations tend not to grow so rapidly at first, but the fastest growing reach a maximum mean annual increment of about 50 m^3/ha/year at around 15–17 years of age and then decline steadily to just over 35 m^3/ha/year at 30 years of age. Some plantations are established at much higher stocking densities, often about 10,000 stems/ha, for production of wood for bioenergy. These may have even higher mean annual increments; values as high as 100 m^3/ha/year to 3 years of age have been reported (Sims et al. 1999).

The general situation for growth of uneven-aged stands is not greatly different from that of even-aged stands. If the development of an uneven-aged stand is followed starting from bare ground, its growth trajectory will usually follow a sigmoid growth pattern as in Fig. 8.4a. However, later in their life cycle, the growth behaviour of uneven-aged stands can differ from that of even-aged stands.

An example of this is given by Moser (1972), who studied the growth over 18 years of uneven-aged stands of a mixture of hardwood tree species native to Wisconsin in the USA. Moser found that their current annual increment in stand basal area was virtually constant, at about 0.15 m^2/ha/year for the entire 18 years. It is clear from Moser's work that the forest with which he was dealing had reached a more or less equilibrium stage of development. Each year some trees died. Each year some new seedlings regenerated below the existing canopy to grow slowly in the shaded environment until they reached the upper canopy many years later. The remaining trees, which were already in the upper canopy, continued to grow year by year until, eventually, they too would be amongst the annual deaths.

Most uneven-aged forests reach this equilibrium stage at the end of the successional development stages through which they pass during their life cycle. This final stage may last for many decades or even hundreds of years until some catastrophic disturbance (a severe fire, a major storm or logging) destroys the forest completely. The forest then has to start its life cycle afresh from bare ground with rapid early growth of species tolerant of the full light conditions, followed by development of species that are able to grow under shade; sometimes this succession process is altered by smaller disturbances that are insufficient to destroy totally the forest. This is the life cycle enjoyed also by rainforests, as discussed in Sect. 8.3. The review by Porté and Bartelink (2002) considers how growth behaviour of uneven-aged forests is modelled. They show a number of examples of how their growth behaviour changes as they pass through their various successional stages.

Even-aged forest stands may reach a stage late in their life when they have zero current annual increment or indeed even a negative current annual increment as some trees die. However, this is only a period of old age and is not accompanied by the continuous cycle of regeneration, death and growth that occurs in the equilibrium stages of uneven-aged forest.

Chapter 9
Measuring Populations

Abstract This chapter introduces the concepts involved in measuring characteristics of whole forest areas, a practice known as forest inventory. Because forests are usually large and complex, it is impossible to measure every individual within them. Instead, a sample only of individuals is measured, and it is assumed the sample represents reasonably the properties of the whole forest population from which the sample was drawn. Usually, the average of the measurements from the sample is used as an estimate of the average of the entire population. Because of the uncertainty involved in this assumption, a measure of the level of uncertainty, known as the confidence limit about the estimate, is calculated also from the sample data. The principles involved in this process are discussed in this chapter.

9.1 Forest Inventory and Sampling

Discussion to this point in the book has dealt with the measurement of individual trees and the measurement of groups of trees (stands). The next four chapters will consider the last stage in scaling up of measurements (Sect. 1.3), that is, inventory of forests. This involves the measurement of whole forests or (as they might be called more formally) forest **populations**.

What is defined as a whole forest (forest population) is completely in the eye of the beholder. To a farmer, it might be an area of only a few hectares of plantation forest on his or her property. To a large company, it might be thousands of hectares of both plantation and native forests supplying many thousands of cubic metres of wood annually to large wood processing plants.

There are many things an owner might want to know about the forest; these might include the availability of wood from it, the occurrence of plant and animal ecosystems within it or the way in which the public uses it. Whatever the owner wants to know, it is obviously impossible, even for quite small forest areas, to measure every tree or every animal in it or the activities of every person who uses it.

To make tractable the problem of measuring whole forests, we resort to sampling. That is, small parts of the forest are measured, usually a tiny fraction of the whole, and it is assumed that the parts measured represent adequately the whole.

The measurements made on the sample are then used to make an estimate of the characteristics of the whole forest.

Sampling is not restricted to forestry. It is used to learn things about populations in all areas of human endeavour. A population can be defined as any set of things about which it is desired to know something. Populations can be big or small. Astronomers might want to know things about the stars in the Milky Way galaxy and will define them as their population; in that case it is a very large population, both in numbers and the space it occupies. Geographers might wish to know things about the people who live in Siberia; those people will constitute their population. An ecologist might wish to learn about the behaviour of ants in the root ball of a single tree that has blown over, and those ants will constitute his or her population. The most important thing about defining a population is to do so clearly. Then, it will be quite apparent what is and what is not included in it and so what things within it are eligible to be included in any sample taken from it.

Speaking formally in mathematical statistical terms, populations are made up of **sampling unit**s. A sampling unit is any clearly defined part of, or individual in, a population and that might be included as one member of a sample drawn from it. In the Milky Way galaxy population, individual stars might be the sampling units. In Siberia, individual people might be the sampling units. In ants in a root ball, individual ants might be the sampling units.

There are many ways both in which a sample can be drawn from a population and how the information derived from that sample is used to learn something about the whole population. Knowing how to sample populations is the realm of sampling theory in mathematical statistics, a complex field of study in its own right. In this book, only a tiny part of that field will be considered.

This chapter considers some basic mathematical statistical issues and methods in Sects. 9.2 and 9.3. In Sect. 9.4, these methods are applied to an example that uses the very simplest sampling method. More complex sampling methods are discussed in Chap. 10.

9.2   Subjective Versus Objective Sample Selection

The selection of a sample from a population could be done subjectively. That is, the sample selector could choose sampling units from the population that he or she considers 'typical'. Or perhaps he or she could select samples at extreme ends of the population and assume that their average represents the population.

However, subjective selection always risks the biases of the selector's judgement; there is no guarantee that the results from such samples will reflect properly the true characteristics of the population. Even worse, if the selector is corrupt, he or she might choose a sample that leads to a population estimate that the selector knows is desired.

In science in general, and forest inventory is no different, sample selection is done objectively so the personal prejudices or fallibilities of the selector play no part. This is usually (although not always) done by a random selection process: given the individuals in a population, tables of random numbers or a computer random number generator is used to select which sampling units will be included in the sample. Modern computer systems have random number generators included as part of them.

In mathematical statistical terms, it is impossible generally to develop theory appropriate for other than objectively chosen samples. Furthermore, if the information learnt about a population from a sample is to be defensible, it must have the imprimatur of mathematical statistical rigour. Thus, in the discussion of sampling for forest inventory, this book will be concerned with the variety of objective strategies that are used for sample selection.

9.3   Population Statistics

Much of what is done in science involves making generalisations about things. Questions are asked like 'how much weight does an ant carry?' or (in a rather less scientific question) 'how far will a family travel on a hot Sunday afternoon to treat the children to an ice cream?' Individual ants vary in how much weight each can carry (ants vary in strength just like people do), and families vary in just how far they will travel for an ice cream (parents vary in their tolerance of their children's demands, and children vary in their passion for ice cream). In science, the objective is usually to make a general statement about such things and avoid the fine detail of the variation amongst individuals.

To make these generalisations, various **population statistics** are used. These are measures used to summarise the properties of populations. Several of importance to forest inventory are described below.

9.3.1   Measures of Central Tendency

Measures of central tendency attempt to summarise the magnitude of whatever it is that is being measured in a population. The measure used most commonly is the average, or **mean** as it is called in mathematical statistics. Then, a generalisation about ants could be made by saying something like 'on average, ants carry a load of 10 mg (milligrams)'. A generalisation about ice-cream-seeking families might be 'the average family will not bother to go out for ice cream if the trip involves a drive of more than 3 km or that takes more than 20 min', information that would be useful to an ice-cream entrepreneur considering where to set up new ice-cream stalls.

Other measures of central tendency are used for various purposes, particularly the **median** (the value in a set of data that has an equal number of values greater than and less than it) and the **mode** (the most common value in a data set). These can be particularly useful when the values in a data set are not spread similarly above and below the mean (i.e. when the data have a skewed distribution).

9.3.2   *Variance and Confidence Limits*

It is of interest (at least to an ecologist) to know that ants carry 10 mg in weight on average. That information will have been derived from a sample of ants in the ant population, a sample obtained by stealing loads from some ants and weighing them. However, we cannot be sure just how representative the sample is of the whole population of ants. Maybe, over the whole population, the ants actually carry average loads of 15 mg, but the average in the sample was only 10 mg. In other words, it is necessary to recognise that any sample taken from a population can only give an estimate of the true value of the measure of central tendency that is being determined for the population.

Mathematical statistics provides an important population statistic, the **confidence limit**, that allows us to state exactly how confident we are that a mean obtained from a sample truly represents the actual mean of the population. It is a measure of the precision of the estimate of the population, where 'the repeated measurements or estimates of something' (see Sect. 2.4) are the measurements taken in the sample, the 'something' in this case being the population.

In turn, the confidence limit is based on a measure of precision called variance that was mentioned also in Sect. 2.4 and was left to be considered further here. A large part of the entire discipline of mathematical statistics is concerned with how to deal with variation in natural systems and, so, with the determination of their variance.

9.4   Calculating the Population Statistics

An example will be used to illustrate the calculation and interpretation of the mean, variance and confidence limit of a population. The population to be considered will be the trees growing in a 0.25 ha plot in an area of native eucalypt forest in northern New South Wales, Australia, a population measured for many years by my forest measurement students. There are 107 trees in this population. The diameter at breast height over bark and stem wood volume of each are given in Table 9.1. The sampling units in the population will be the individual trees. Their volumes vary over the range 0.013–1.977 m^3 and their average volume is 0.424 m^3. As populations go, this is a very small and simple population. But, it will serve perfectly well to illustrate how we conduct an inventory.

Suppose the stem wood volumes of all the trees in this population had not actually been measured and it was desired to estimate their mean volume by selecting a sample from them. The stem wood volumes only of the trees selected

Table 9.1 The diameter at breast height over bark (DBH) and stem wood volume of each of a population of 107 trees in a eucalypt forest plot

Tree	DBH (cm)	Volume (m^3)	Tree	DBH (cm)	Volume (m^3)	Tree	DBH (cm)	Volume (m^3)
1	46.5	1.977	37	22.7	0.520	73	20.7	0.154
2	42.0	1.529	38	25.0	0.495	74	15.6	0.143
3	41.4	1.514	39	26.5	0.489	75	16.5	0.142
4	40.0	1.457	40	24.2	0.484	76	18.6	0.141
5	41.5	1.312	41	24.5	0.449	77	14.3	0.132
6	35.5	1.194	42	21.0	0.422	78	16.5	0.125
7	36.5	1.158	43	24.4	0.414	79	14.5	0.124
8	37.2	1.145	44	26.3	0.382	80	15.0	0.118
9	34.0	1.074	45	21.0	0.369	81	12.0	0.116
10	35.2	0.993	46	26.3	0.336	82	11.6	0.111
11	34.3	0.958	47	21.2	0.334	83	13.5	0.111
12	32.7	0.939	48	22.5	0.333	84	13.7	0.110
13	32.5	0.913	49	22.2	0.332	85	15.6	0.105
14	33.3	0.901	50	19.5	0.324	86	11.5	0.102
15	31.8	0.851	51	20.8	0.323	87	10.3	0.101
16	29.6	0.789	52	19.5	0.320	88	12.6	0.101
17	28.0	0.731	53	18.8	0.316	89	15.0	0.097
18	28.7	0.726	54	20.0	0.301	90	8.5	0.093
19	30.0	0.722	55	21.8	0.301	91	14.5	0.088
20	30.0	0.717	56	20.2	0.272	92	16.0	0.088
21	28.8	0.707	57	19.5	0.271	93	13.5	0.082
22	30.5	0.690	58	18.9	0.268	94	13.0	0.073
23	26.5	0.680	59	17.7	0.254	95	14.6	0.072
24	30.0	0.675	60	20.3	0.249	96	12.5	0.063
25	28.0	0.672	61	19.5	0.246	97	11.5	0.060
26	27.9	0.662	62	13.9	0.235	98	12.0	0.059
27	28.0	0.642	63	22.0	0.212	99	12.2	0.059
28	23.6	0.641	64	16.2	0.199	100	9.0	0.056
29	27.7	0.627	65	13.0	0.191	101	16.5	0.053
30	24.5	0.565	66	14.5	0.186	102	16.6	0.046
31	27.3	0.562	67	21.0	0.186	103	14.5	0.045
32	27.0	0.552	68	18.0	0.183	104	13.5	0.032
33	25.6	0.546	69	16.3	0.176	105	16.3	0.032
34	23.5	0.536	70	17.7	0.173	106	13.6	0.026
35	27.3	0.530	71	16.5	0.166	107	11.2	0.013
36	25.3	0.528	72	16.0	0.154			

The average volume of the trees in the plot is 0.424 m^3

in the sample would then be measured. Suppose it was decided to select 15 trees from the population as the sample and, to ensure objectivity, these trees were selected randomly. Suppose also that the sampling was done without replacement

Table 9.2 Sample of trees
selected by simple random
sampling from the population
of trees in Table 9.1

Tree	DBH (cm)	Volume (m³)
4	40.0	1.457
20	30.0	0.717
29	27.7	0.627
41	24.5	0.449
49	22.2	0.332
67	21.0	0.186
75	16.5	0.142
85	15.6	0.105
90	8.5	0.093
96	12.5	0.063
100	9.0	0.056
102	16.6	0.046
105	16.3	0.032
106	13.6	0.026
107	11.2	0.013

(i.e. once a tree was selected in the sample, it was not considered for inclusion again); most sampling in forest inventory is done without replacement.

Table 9.2 lists a sample of 15 trees selected randomly from the population (the selection was done with the aid of a random number generator on my computer). After selecting the sample, the wood volumes of the stems of those trees would have been measured using one of the techniques discussed in Chaps. 5 and 6.

Suppose there are n individuals in the sample ($n = 15$ in this case) and the stem wood volumes measured on those 15 trees are denoted as y_i (with i taking the values 1, 2...n), so that $y_1 = 1.457$, $y_2 = 0.717$ ... $y_{15} = 0.013$ (as in Table 9.2). Using the sample data, an estimate of the mean volume of the trees in the population, $\overline{Y}$, is calculated as the average of the sample, that is,

$$\overline{Y} = \left[\Sigma_{i=1...n} y_i \right] / n. \tag{9.1}$$

For the sample, this gives an estimate of the mean volume of the trees of the population of 0.290 m³, a good deal less than the actual mean of 0.424 m³ for the whole population. This emphasises how far the estimate of the population mean determined from a sample can deviate from the true mean of the population.

An estimate of the variance of stem wood volume in the population, $V(\overline{Y})$, is then determined from the sample data as

$$V(\overline{Y}) = \left[\Sigma_{i=1...n} (y_i - \overline{Y})^2 \right] / [n - 1]. \tag{9.2}$$

For the example, $V(\overline{Y}) = 0.155$. The confidence limit about the estimate of the population mean, $C(\overline{Y})$, is then given by

$$C(\overline{Y}) = t \sqrt{[V(\overline{Y})/n]} \tag{9.3}$$

where t is the value of a variable known by the name Student's t. The variable t was discovered in 1908 by W. S. Gosset, who published under the pseudonym 'Student'. Its value varies, depending both on how confident we wish to be of our final result and on the number of observations in our sample. Suppose we choose to be 95 % confident of our result and, given there are $(n-1) = 14$ degrees of freedom, as they are called, in our sample, then the value of t (that can be looked up in a table) is 2.145. So, the value of $C(\overline{Y})$ for the example is $2.145 \times \sqrt{(0.155/15)} = 0.218$ m^3.

These results from the sample are interpreted by saying that we can be 95 % confident that the actual mean of the population lies within ± 0.218 m^3 of the estimate of the population mean obtained from the sample (0.290 m^3). That is, we can say we are 95 % confident that the population mean lies within the range 0.290 ± 0.218 m^3, that is, within the range 0.072–0.508 m^3. This range is called the **confidence interval** of the estimate of the population mean. In this example, the actual population mean, 0.424 m^3, does indeed lie within that range; because a confidence level of 95 % was chosen, it would be expected that in 5 % $(100 - 95\,\%)$ of samples taken from the population, their confidence intervals would not include the population mean. The confidence limit is our measure of the precision of the estimate of the population mean.

The determination of a confidence limit (hence, confidence interval) about an estimate of a population mean is an extremely important part of an inventory of any population. It is used to indicate how worthwhile the inventory has been. If the confidence limit is low, relative to the estimate of the population mean, it can be said that the characteristics of the population have been estimated well. If it is high, it can be said that the characteristics have been estimated poorly.

In the example, the confidence limit is rather high in relation to the estimate of the population mean (it is 75 % of the estimate of the mean). If it was desired seriously to know the average tree stem wood volume in the example population, and the best that could be said was that it lay somewhere within the range 0.072–0.508 m^3, then we would probably not be very happy with the result. However, it is most important to recognise that it is entirely up to the person for whom the estimate is being made to judge whether or not the width of the confidence interval is adequate for the purpose for which the estimate is required.

If the confidence interval is wider than desired, there are two options available. The first is to increase the size of the sample. In our example, when a new random sample of 30 trees was taken from the population, the estimate of the mean was 0.349 m^3 and of the 95 % confidence interval was 0.189–0.509 m^3. This is still quite a wide interval, but much narrower than the range of 0.072–0.508 m^3 that was found with the sample of 15 trees. The second option is to adopt a different strategy to select the sample; options that might be used in forest inventory to do this are discussed in Chap. 10.

It should be noted also that it is entirely up to the judgement of the person for whom the inventory is being done as to what level of confidence should be used in determining the confidence limit of the estimate. Quite arbitrarily in the example, it was chosen that we should be 95 % confident about the result. That is a common level of confidence used in the biological sciences. For a forest inventory, the person for whom it is being done might be happy with a result of which he or she was only 90 % confident, or he or she might demand a result of which they can be 99 % confident.

The value of Student's t, that is used in (9.3), varies with the level of confidence chosen, as well as with the number of observations in the sample. So, if it was desired to be only 90 % confident of the result in the example, the value of t to be used in (9.3) would be 1.761, or 2.977 if it was desired to be 99 % confident. These values are smaller and larger, respectively, than the value of 2.145 that was used to be 95 % confident. The width of the corresponding confidence intervals would be progressively wider the more and more confident of the result it was desired to be. Of course, it is impossible to ever be 100 % confident of the result from a sample; the confidence interval would be infinitely wide in that case. To be 100 % confident, each and every individual in the entire population would have to be measured. Tables with values of Student's t are provided commonly in statistics textbooks, and computer packages often provide functions with which they can be calculated.

Chapter 10
Sampling Theory

Abstract Estimating forest characteristics over large areas in forest inventory is a complex and expensive task. It is important to achieve the estimates at minimum cost but with no more uncertainty than is appropriate for the objectives of the inventory. Forest inventory involves taking samples randomly from across the whole forest population and measuring the forest characteristics on the selected sample. There are many ways of selecting which parts of the forest are to be included in the sample. Further, some sampling techniques provide estimates of forest characteristics with less uncertainty than others, although always with additional measurement effort. This chapter describes a number of different sampling techniques that are used in forest inventory.

10.1 Sampling Techniques and Their Efficiency

As mentioned in Sect. 9.4, there are various techniques that can be adopted to select a sample from a population. In the example used in Chap. 9, the sample (Table 9.2) was selected at random from the population (Table 9.1) in such a way that each and every tree in the population was equally likely to have been included in the sample. Speaking in mathematical statistical terms, this is a simple random sample (often abbreviated as SRS), that is, a sample in which each and every sampling unit in the population has the same probability of selection (or in common parlance, the same chance of being selected). In the example, there were 107 trees in the population and 15 trees were to be sampled. Thus, the probability of selection of any of the 107 trees was 15/107, that is, 0.140, or a 14 % chance.

Simple random sampling is the easiest technique by which a sample might be selected, but it is certainly not the only one. Importantly, there are other sampling techniques that have the great advantage that they may lead to a reduction in the size of the confidence limit of whatever it is that is being estimated about a population. Sampling forests can be a time-consuming and expensive task. So, it is obviously desirable to obtain the most precise estimate possible of the characteristics of the forest population with the least sampling effort. Again speaking formally, one sampling technique that leads to a more precise population estimate

© Springer International Publishing Switzerland 2015
P.W. West, *Tree and Forest Measurement*, 3rd edition,
DOI 10.1007/978-3-319-14708-6_10

(i.e. an estimate with a lower confidence limit) than another is said to be a more efficient technique.

In forest inventory practice, these more efficient sampling techniques rely on the availability from the population of what is known as '**auxiliary**' **information** (known also as '**covariate**' or '**ancillary**' **information**). Auxiliary information involves one or more characteristics of the forest that are correlated, at least partially, with the variable of interest that it is actually being measured in the population. This auxiliary information may have been gathered from the forest before the inventory was to be carried out. Otherwise, it might be gathered as part of the inventory, in which case it is usually information that can be obtained readily and widely across the population and does not require intensive field measurements by ground crews. In modern forest inventory, it is common to obtain auxiliary information widely across the forest by remote sensing using sophisticated measuring equipment borne aloft in aircraft or from satellites; Gräfstrom and Ringvall (2013) discussed the use of such information in developing appropriate sampling techniques for forest inventory.

This chapter will describe several sampling techniques that are more efficient than simple random sampling but require that auxiliary information from the forest be available.

10.2 Stratified Random Sampling

The first more efficient sampling technique considered uses auxiliary information to group together sampling units that have similar auxiliary information values. These groups, or strata as they are called, are chosen so that the variation between the sampling units in any one group (**stratum**) is less than the variation over the whole population. This process is called stratified random sampling.

Consider a large area of a particular type of forest (such as a plantation forest or a rainforest). Across the area, the forest will vary in age or stage of development, as will the topography, climate and soil type. All of these may be considered as auxiliary information and will affect the amount of wood or biomass (or of many other variables that might be of interest in a forest inventory) present at any point in the forest. So, stratification of the forest area into smaller areas based on these auxiliary characteristics would be expected to reduce the variation in any one of the strata when compared with the variation over the whole population. The more and more the variation between sampling units within any one stratum can be reduced, the more efficient will stratified random sampling be.

In stratified random sampling, the auxiliary information may be categorical or numerical. Thus, different soil types might constitute different categories on which to base a stratification. As well, information from several different types of auxiliary variable might be used. Thus, strata defined by soil type might be subdivided further into strata defined also by age of the forest, age constituting auxiliary information that takes a numerical value, rather than categorical.

Once the strata have been defined, the computations required with stratified random sampling proceed as follows. Suppose there are H strata. Suppose the hth stratum $(h = 1 \ldots H)$ contains $N(h)$ sampling units in total and $n(h)$ of those are included in a simple random sample selected from that stratum. Suppose a value of the variable of interest to the inventory, $y(h,i)$, was measured on the ith $[i = 1 \ldots n(h)]$ sampling unit that was sampled in the hth stratum. The estimate of the mean of the variable of interest in the hth stratum, $\bar{y}(h)$, is then given by

$$\bar{y}(h) = \left[\Sigma_{i=1 \ldots n(h)} y(h,i) \right] / n(h) \tag{10.1}$$

Following Eq. 3.14 of Schreuder et al. (1993) with slight modification to estimate the mean rather than total, the estimate of the population mean, $\bar{Y}$, from the stratified random sample is then

$$\bar{Y} = \left\{ \Sigma_{h=1 \ldots H} \left[W(h) \bar{y}(h) \right] \right\}, \tag{10.2}$$

where $W(h) = N(h)/N$ and N is the total number of sampling units in the population, that is, $N = \Sigma_{h=1 \ldots H} N(h)$. The estimate of the variance of the estimate of the population mean is (following Eq. 3.16 of Schreuder et al. (1993), with modification also for variance of the mean rather than the population total)

$$V(\bar{Y}) = \Sigma_{h=1 \ldots H} \left[W(h)^2 \{ 1 - f(h) \} \left\{ \Sigma_{i=1 \ldots n(h)} [y(h,i) - \bar{y}(h)]^2 \right\} / \{ [n(h) - 1] n(h) \} \right], \tag{10.3}$$

where $f(h) = n(h)/N(h)$. The confidence limit can then be determined as

$$C(\bar{Y}) = t \sqrt{V(\bar{Y})} \tag{10.4}$$

with the number of degrees of freedom for Student's t being approximately $[\Sigma_{h=1 \ldots H} n(h)] - h$ (Satterthwaite 1946; Cochran 1977; Gregoire and Valentine 2008).

To illustrate how stratified random sampling works, it will be applied to the example from Sect. 9.4. The aim of the inventory in the example was to estimate the average tree stem volume in the population. In the example some auxiliary information is provided, namely, the diameter at breast height over bark of each tree in the population (Table 9.1), a tree characteristic that can be expected to be correlated to some degree with stem wood volume. Suppose the trees in that population were divided into three strata based on their diameters. Suppose the first stratum was made up of trees with diameters in excess of 25.5 cm (there are 35 such trees in the example); this might be called the 'large tree' stratum. Suppose the second stratum (the 'medium tree' stratum) contained trees with diameters in the range 16.5–25.5 cm (37 trees). Finally, suppose the third (the 'small tree' stratum) contained trees with diameters less than 16.5 cm (35 trees). Thus, in our example $H = 3$, $N = 107$, $N(1) = 35$, $N(2) = 37$ and $N(3) = 35$.

Since tree diameter is expected to be correlated to some extent with stem volume (trees with larger stem diameters can be expected to have larger volumes), it would be expected that the variation in stem volume in any one of these three strata would be less than the variation over the whole population. In reality, this is a most unlikely form of stratification to do. Where numerical auxiliary information like diameter is available for the sampling units in a population, we are more likely to use one or other of the sampling techniques discussed in succeeding sections of this chapter. Those techniques can be expected to make much better use of that auxiliary information and would be expected to be more efficient sampling techniques than stratified random sampling. However, for the sake of the example, we will persist with those three strata here.

Suppose a sample of 15 trees was now selected at random from the population. In doing so, the number of trees selected from each stratum will be approximately proportional to the stratum size. The simple random sample selected earlier (Table 9.2) will serve for this purpose. From the tree diameters in that table, it will be seen that this sample includes three trees from the large tree stratum, five from the medium tree stratum and seven from the small tree stratum. Thus, for the example $n(1) = 3$, $n(2) = 5$ and $n(3) = 7$. The values of $y(h,i)$ are then the measured tree stem volumes in Table 9.2 for the sample trees in the respective strata. The results of applying (10.1–10.4) to these data are shown in the second row of Table 10.1. Comparison of this confidence interval with that from simple random sampling (as calculated in Sect. 9.4 and shown in the first row of Table 10.1) shows it is appreciably narrower for stratified random sampling. That is, stratified random sampling has indeed been a more efficient technique than simple random sampling.

One problem with this theory for stratified random sampling is that it requires that the number of sampling units in each stratum [the $N(h)$] be known. In large and complex forest populations, these are usually unknown. Furthermore, if point sampling (Sect. 8.4.2) is being used to take the sample measurements, there is an infinite number of points at which the samples could be taken in any stratum.

In practical forest inventory, this problem can usually be overcome readily. Most commonly, stratification involves using the auxiliary information to map out the stratum areas across the population. Thus, if stratification was being done on the

Table 10.1 For various sampling techniques, estimates of the population mean stem wood volume (m^3) and its 95 % confidence interval for the tree population of Table 9.1

Sampling technique	Mean	95 % confidence interval
Simple random sampling	0.290	0.072–0.508
Stratified random sampling	0.405	0.220–0.590
Sampling with probability proportional to size	0.413	0.365–0.461
Sampling with probability proportional to prediction	0.407	0.352–0.462
Model-based sampling	0.390	0.342–0.438

The true mean stem wood volume of the population was 0.424 m^3

basis of soil type and age, say, areas of any particular age and soil type would be mapped out and the area of each of those strata, $A(h)$ ($h = 1...H$), would be determined. Then, the values of $W(h)$ in (10.2) and (10.3) would be calculated as $W(h) = A(h)/A$, where A is the total forest area [$A = \Sigma_{h=1...H} A(h)$]. As well, in most practical forestry inventories (our very simple example is a case where this does not apply), the total number of sampling units in any one stratum [$N(h)$] is much greater than the sample size selected from that stratum [$n(h)$], that is, $N(h) \gg n(h)$. When this is so, $f(h) = n(h)/N(h) \approx 0$ and so a value of zero may be used for $f(h)$, for all h, in (10.3). With these modifications, the values of $N(h)$ no longer appear in (10.2) or (10.3) and so the theory can be applied readily in practice.

10.3   Sampling with Probability Proportional to Size

Sampling with probability proportional to size (often referred to as PPS sampling) may be applied when numerical (not categorical) values are available in advance for an auxiliary variable for all, or at least a large number of, the sampling units right across the entire population. In modern forest inventory, such auxiliary information is often obtained by remote sensing with aerial- or satellite-borne equipment.

In PPS sampling, the auxiliary variable must be correlated positively with the variable of interest. Having a positive correlation means that sampling units with a larger value of the auxiliary variable tend to have a larger value of the variable of interest (a negative correlation would mean that a larger value of the auxiliary variable would tend to be associated with a smaller value of the variable).

Suppose that the entire population consisted of N sampling units and the value of the auxiliary variable measured on the ith ($i = 1...N$) sampling unit was x_i. In PPS sampling, the probability of inclusion of each sampling unit in the sample is no longer assumed to be the same for each sampling unit as it was with simple random sampling (Sect. 10.1). Instead it is determined from the auxiliary variable values so that the probability of inclusion of the ith sampling unit, p_i, is

$$p_i = n\left[x_i / \left(\Sigma_{i=1...N} x_i\right)\right], \tag{10.5}$$

where n is the size of the sample that it is desired to select from the population.

Consider the example from Sect. 9.4. The auxiliary variable measured on each tree in the population was its stem diameter at breast height over bark. We know from Chap. 6 that tree diameter is likely to be correlated positively with tree stem wood volume and that the square of tree diameter (in effect, tree basal area) is likely to be particularly well related to volume. Thus, the square of tree diameter might be a highly appropriate auxiliary variable for this example. Given that $N = 107$, we would then determine auxiliary variable values (see the data in Table 9.1) of $x_1 = 46.52^2 = 2,162.25$, $x_2 = 42.0^2 = 1,764.00 \ldots x_{107} = 11.2^2 = 125.44$.

Suppose it is now desired to select a sample of $n = 15$ trees from the population by PPS sampling. Given that $\Sigma_{i=1...N} x_i = 58,939.42$, then using (10.5) a probability of selection could be assigned to each tree as $p_1 = 15 \times 2{,}162.25/58{,}939.42 = 0.55$, $p_2 = 0.45 \ldots p_{107} = 0.03$. Thus, tree 1 would have a 55 % chance of being included in the sample, tree 2 a 45 % chance and tree 107 only a 3 % chance; it is obvious from (10.5) that PPS sampling favours inclusion in the sample of trees with larger values of the auxiliary variable. It can be proved mathematically that this is likely to lead to more efficient sampling than simple random sampling.

To decide which sampling units are then to be actually included in the sample, the sampling units are considered in a random order and a randomly chosen value in the range 0–1 is assigned to each; computer systems provide functions that give such random values. If the probability of selection assigned to that sampling unit exceeds the random value chosen for it, that sampling unit is then included in the sample. This process continues until the required sample size is reached. Table 10.2 lists a sample of 15 trees that were selected using this process. Study of that table will show that it does include a higher proportion from the population of trees of larger diameter than trees of smaller diameter.

Once the sample has been selected, the tree stem wood volumes would be measured of the sample trees and these are shown in Table 10.2. Suppose the jth tree in the sample had a stem wood volume of y_j, with a corresponding probability of inclusion in the sample of p_j. Following Eqs. (3.7) and (3.9) of Schreuder et al. (1993) (with some algebraic rearrangement), the estimate of the population mean stem wood volume would then be

$$\overline{Y} = \left[\Sigma_{j=1\ldots n} \left(y_j/p_j \right) \right]/N \qquad (10.6)$$

and of its variance would be

Tree	DBH (cm)	Volume (m³)
3	41.4	1.514
5	41.5	1.312
6	35.5	1.194
10	35.2	0.993
17	28.0	0.731
20	30.0	0.717
29	27.7	0.627
31	27.3	0.562
33	25.6	0.546
35	27.3	0.530
38	25.0	0.495
45	21.0	0.369
73	20.7	0.154
74	15.6	0.143
89	15.0	0.097

Table 10.2 Sample of trees, selected by sampling with probability proportional to size (PPS sampling), from the population of trees in Table 9.1

$$V\left(\overline{Y}\right) = [N - n]\left[\Sigma_{j,k=1...n}\left(y_j/p_j - y_k/p_k\right)^2\right] \Big/ \left[2N^3(n - 1)\right].   \qquad (10.7)$$

Note that in (10.7) the mathematical expression $\Sigma_{j,k=1...n}$ denotes the summation of the term that follows the expression, as both j and k take successive values 1, 2, 3... up to n. Equation (10.4) would then be used to determine the confidence limit of the estimate of the mean with $n - 1$ degrees of freedom for Student's t.

The results of applying this process to the example are given in the third row of Table 10.1. As anticipated, it gave a confidence interval that was appreciably narrower than that from simple random sampling. It was also appreciably narrower than for stratified random sampling which, as discussed in Sect. 10.2, was unable to make as good use of the numerical auxiliary information as PPS sampling.

In large forest populations, the total number of sampling units in the population will often be unknown and/or it will not be possible to obtain an auxiliary variable value for each and every one. Fortunately, all that is required for PPS sampling is that auxiliary variable values be obtained for a relatively large number of randomly selected sampling units in the population, a number large enough that they can reasonably be assumed to represent the entire population, as if the entire population was indeed being considered. Usually a few thousand sampling units will be more than sufficient for this purpose and often many less. The theory described above is then applied as if this large set of sampling units was the entire population; that is, the value of N used is the size of this set of sampling units. In modern forest inventory, where auxiliary variable values are often obtained using airborne or satellite equipment, it is common to gather easily a very large number of auxiliary variable values from widely across the entire population.

10.4 Sampling with Probability Proportional to Prediction

Before a sample was selected using PPS sampling (Sect. 10.3), it was necessary to have available a value of an auxiliary variable for a large number of sampling units over the entire population. As mentioned earlier, imagery from airborne or satellite equipment is often used to provide this information in modern forest inventory.

This type of imagery is both expensive to obtain and requires specialist staff to process the raw information and make it available in a form that can be used for forest inventory. It is common for much smaller-scale forest inventories to be carried out, where the cost of such auxiliary information may be too high for the inventory budget or there is insufficient time and resources available to collate such information. Also, it may simply be that the auxiliary variable cannot be measured from aloft using this sophisticated equipment, but must be measured on the ground. In any of these circumstances, sampling with probability proportional to prediction (often abbreviated as 3P sampling and sometimes called Poisson sampling) may offer the advantages of the efficiency of PPS sampling but without the need to have the auxiliary variable values available in advance widely across the population. Instead, in this form of sampling, auxiliary variable values need be determined only for those sampling units that are considered for inclusion in the sample and only at the time the sampling is being done in the field. Just as with PPS sampling, the auxiliary information must involve numerical values and must be correlated positively with the variable of interest to the inventory.

This sampling technique was invented in the 1970s by an American forest scientist, L. R. Grosenbaugh. It has been used quite extensively in America. However, in the form in which it was developed originally, its use has been restricted largely to very precise estimation of the availability of wood volume across rather small forest tracts, usually only tens of hectares in size. West (2011) discussed its possible more general application to forest (or natural resource) inventory over large areas. The form of 3P sampling he described was termed 'ordinary' 3P sampling and it is that form of sampling that will be considered here.

The first step in 3P sampling (ordinary or otherwise) requires that a preliminary survey of the population be carried out to determine the maximum and minimum values of the auxiliary variable that occur anywhere across the population. Suppose they are found to be x_u and x_l, respectively. It is important that these values be determined carefully. If, during sampling, a sampling unit is encountered with an auxiliary variable value outside this range, the inventory must be restarted.

Once values of x_u and x_l have been determined, sampling may start. A randomly selected sampling unit within the population is visited and its (quickly and easily measured) auxiliary variable value determined. A random value is then selected from within the range x_l–x_u. If the auxiliary variable value measured is equal to or greater than this randomly selected value, the variable of interest to the inventory is measured on that sampling unit. If not, the sampler simply moves on to the next randomly selected sampling unit. This is what is known as a two-phase sampling process. The first phase consists of the randomly selected sampling units (they constitute a simple random sample of the population) on all of which the auxiliary variable value was measured. The second phase selects a 3P subsample of the first-phase sampling units on which the variable of interest to the inventory is measured also.

This sampling process continues until the second-phase sample reaches some desired sample size of n sampling units. By that time, a total of f sampling units ($f \geq n$) will have been visited in the first-phase sample. Suppose the ith member of the first-phase sample has an auxiliary variable value of x_i. If this is chosen as the jth member of the second-phase sample, its auxiliary variable value can then be denoted as x_j and its value of the variable of interest to the inventory as y_j.

Table 10.3 shows a sample selected by this process from the example population of Table 9.1. As for PPS sampling, the square of tree diameter was used as the auxiliary variable value in this process. Before starting the sampling process and after inspection of the population data, it was assumed the auxiliary variable values lay within the range 50–2,500 cm^2, that is, this was the range used for x_l–x_u. Individual sampling units (trees) in the population were then considered in random order and the 3P sampling process applied. In this case, $f = 55$ sampling units were

Table 10.3 Sample of trees, selected in the first-phase sample by sampling with probability proportional to prediction (3P sampling), from the population of trees in Table 9.1

Tree	DBH (cm)	Volume (m^3)	Tree	DBH (cm)	Volume (m^3)
1	46.5	1.977	59	17.7	
4	40.0		62	13.9	
7	36.5	1.158	63	22.0	
9	34.0	1.074	65	13.0	
10	35.2	0.993	66	14.5	
12	32.7	0.939	68	18.0	0.183
13	32.5		69	16.3	
16	29.6	0.789	70	17.7	0.173
18	28.7	0.726	71	16.5	0.166
21	28.8	0.707	76	18.6	
24	30.0		77	14.3	
30	24.5		78	16.5	0.125
31	27.3	0.562	79	14.5	
33	25.6	0.546	81	12.0	
36	25.3		82	11.6	
37	22.7	0.520	84	13.7	
38	25.0		85	15.6	
40	24.2		86	11.5	
41	24.5		87	10.3	
42	21.0		89	15.0	
43	24.4		94	13.0	
44	26.3		98	12.0	
45	21.0		99	12.2	
46	26.3		101	16.5	
47	21.2		103	14.5	
48	22.5		104	13.5	
56	20.2		105	16.3	
58	18.9				

The stem volumes (the variable of interest in this example) are shown of the trees that were included also in the second-phase 3P subsample

visited before $n = 15$ were included in the 3P subsample, the sample size being used in the examples being considered here; the stem wood volumes of those 15 selected sampling units are shown in Table 10.3. Examination of the table will show that this sampling process has favoured inclusion of trees with larger diameters in the 3P subsample, just as was the case with PPS sampling.

Following Eq. [9] of West (2011), the estimator of the population mean is then determined as

$$\overline{Y} = (1/f)\left\{\left(\Sigma_{i=1...f}x_i\right)\right\}\left\{\Sigma_{j=1...n}y_j/\left(x_j - x_1\right)\right\}/\left\{\Sigma_{j=1...n}x_j/\left(x_j - x_1\right)\right\}. \qquad (10.8)$$

To date, no satisfactory analytical estimator (i.e. a formal mathematically derived equation) has been developed to allow the variance of this estimate of the population mean to be estimated reliably. Instead, West (2011) and some other authors have used a process known as 'bootstrapping' to determine the variance from 3P samples. Bootstrapping uses the data that was collected in the sampling process in what is known as a Monte Carlo or simulation technique. That is, bootstrapping involves averaging many trials that involve random processes; the use of 'Monte Carlo' to describe such techniques derives from the random nature of gambling at the casinos there.

For ordinary 3P sampling, bootstrapping may be applied as follows (West 2011). Given the original first-phase sample of f sampling units, a new sample, also of size f, is selected from the original sample by simple random sampling from it with replacement. That is, when any sampling unit has been selected from the original f into the new sample, it is kept in the list of f and is eligible to be selected again; when this is done, about 37 % of the original sample points will be duplicated in the new sample. The new sample would then be used in exactly the same way as the original sample, using as its 3P subsample whichever sampling units were selected from the original 3P subsample; the size of the new 3P subsample (n) may differ from the size of the original. A new estimate of the population mean would then be determined using (10.8).

This process would be repeated a number of times (experience of the present author suggests 100 times might be sufficient, but many more can be done if desired) with new samples being selected with replacement from the original sample each time. The variance of these new estimates of the population mean (i.e. they would be the values of y_i used in (9.2)) would then be used as the estimate of the variance of the estimate of the population mean $\left[V\left(\overline{Y}\right)\right]$. To determine the confidence limit, this estimate would be used in (10.4) with the number of degrees of freedom for Student's t being $n - 1$.

Bootstrapping has been a controversial technique amongst mathematical statisticians. However, sufficient has been learnt about its properties that it seems to be accepted now as appropriate whenever a more formal alternative is not available. It could be applied to determine confidence limits for any of the sampling techniques considered in this chapter. However, where formal, mathematical estimators of variance have been developed for them, those estimators would be preferred generally to bootstrapping.

The estimate of the population mean from the sample shown in Table 10.3 and its 95 % confidence interval found by bootstrapping are given in the fourth row of Table 10.1. In this case, ordinary 3P sampling has given a 95 % confidence limit of about the same size as PPS sampling. They are not exactly the same size since random effects involved in taking the samples will always ensure that there are differences. However, in practical terms ordinary 3P sampling has given a similar result to PPS sampling because both techniques favour inclusion in the sample of sampling units with larger values of the auxiliary variable in more or less the same fashion.

10.5 Model-Based Sampling

The last sampling technique to be discussed here is called model-based sampling. As with PPS sampling, model-based sampling is appropriate where numerical values of at least one auxiliary variable have been measured initially on a large number of sampling units scattered randomly across the whole population. Importantly however, model-based sampling can be used to even better advantage than PPS sampling if data for more than one auxiliary variable are available.

In model-based sampling, the results from taking a sample from the population are used to establish a regression equation (Sect. 6.2.1) relating the variable of interest to the auxiliary variable(s). The fitted regression equation is then used to predict the values, from the auxiliary variable(s) values, of the variable of interest on all the sampling units on which auxiliary variable values were measured.

There are several advantages with model-based sampling over the other sampling techniques discussed in this chapter. These are:

- Any number of auxiliary variables may be used and that may allow much greater sampling efficiency.
- The auxiliary variables may be correlated positively or negatively with the variable of interest in the population and still be just as useful in model-based sampling; only a positive correlation is useful in PPS or 3P sampling.
- The full power of regression analysis can be brought to bear to establish a relationship between the variable of interest and the one or more auxiliary variables for which information is available in the population. This allows relationships with complex functional forms to be used where these provide the best relationship between the variable of interest and the auxiliary variable(s).
- The way in which the sample is selected from the population need not be as formal as in the other sampling techniques. Texts on regression analysis discuss the optimum sort of information required to fit regression relationships reliably. Suffice to say that the data collected should encompass generally the range of values of the auxiliary variables that occur across the population and that the sampler should have been objective in selecting the sample (i.e. no sampling unit should have been included in the sample through any prejudice of the sampler).

To illustrate model-based sampling, suppose the sample selected earlier for simple random sampling (Table 9.2) was available. Figure 10.1 shows tree stem wood volume plotted against the auxiliary variable, tree diameter at breast height over bark, for the 15 trees in that sample.

Fig. 10.1 Scatter plot of
tree stem wood volume
against tree diameter at
breast height over bark for
the sample selected for
simple random sampling as
shown in Table 9.2. The
solid line shows the
ordinary least-squares
regression fit to the data of
(10.9)

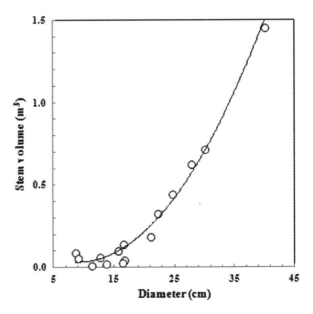

For these data, there appears to be a slightly curved relationship between stem
volume (y_i, m^3) and tree diameter (x_i, cm), where $i = 1\ldots15$ for the 15 observations
in the sample. To fit this curved relationship, it might be appropriate to assume that
there are in fact two auxiliary variables, the second being the square of tree
diameter. Then, an appropriate regression equation to fit to these data might be
what is termed a quadratic relationship (i.e. a relationship that includes a term and
its square), that is, the relationship

$$y_i = a + bx_i + cx_i^2 \tag{10.9}$$

where a, b and c are parameters of the equation. When ordinary least-squares
regression (the form of regression analysis used normally in the biological sciences
and that is discussed in all texts on regression analysis) was used to fit these data to
(10.9), estimates were found for $a = 0.1554$, $b = -0.0276$ and $c = 0.0015$. The fit to
the data of this function is shown as the solid line in Fig. 10.1.

Suppose that the auxiliary variable values had been measured on a total of
N sampling units in the whole population. In the example, N was the entire
107 individual trees in the population (Table 9.1), but may simply be a large
number of randomly selected sampling units as in PPS sampling. The estimate of
the population mean stem wood volume, $\overline{Y}$, is determined by using the fitted
regression function to predict the wood volume of each and every tree in the
population for which the auxiliary variables were measured and adding up the
predicted values, so that

$$\overline{Y} = \left[\Sigma_{j=1...N} \left(0.1554 - 0.0276x_j + 0.0015x_j^2 \right) \right] / N. \qquad (10.10)$$

The result, an estimate of 0.390 m³, is shown in the fifth row of Table 10.1.

Bootstrapping is a method that may be used with this form of sampling to determine the confidence limit of the estimate of the mean. It can be used no matter what form of equation, with however many auxiliary variables, was used to fit the regression equation between the variable of interest and the auxiliary variables. In this case, bootstrapping involves resampling with replacement from the sample data (the data of Table 9.2 in the example here), fitting the regression equation (10.9) with each new sample and using that to find a new estimate of the population mean. After many such new samples have been selected and means estimated, the variance of the estimates of those means may then be used as the estimate of the variance of the estimate of the population mean $\left[V(\overline{Y}) \right]$ and (10.4) used to determine the confidence limit, with the number of degrees of freedom for Student's t being $n - 1$, where n was the sample size ($n = 15$ in the present example).

The confidence interval determined by bootstrapping in the present example is shown in the fifth row of Table 10.1. In this case, model-based sampling appears to have been the most efficient of all the sampling techniques tested here.

An example of the use of model-based sampling in practice can be found in Hamilton and Brack (1999) who used it to estimate wood availability from a large area of native regrowth eucalypt forest in Victoria, Australia. They obtained auxiliary information for their forest from aerial photographs (Sect. 13.2.1) as well as from mapped topographic and climatic data. Stephens et al. (2012) used model-based sampling to estimate the amount of carbon contained in the biomass of forests in New Zealand. They used an airborne laser scanning device (Sect. 13.2.2) to obtain auxiliary information that reflected stand biomass.

10.6   Choosing the Sampling Technique

The discussion in the preceding sections of the various sampling techniques should have made it apparent when each is appropriate. Firstly, if no auxiliary information is available about the population, there is no option but to use simple random sampling.

The technique that then requires the least auxiliary information is sampling with probability proportional to prediction (3P sampling). To apply it, an initial survey must be undertaken to determine the largest and smallest values that occur in the population of the auxiliary variable. The better this maximum and minimum are determined, the fewer sampling units will have to be visited to find those to be included in the second-phase 3P subsample.

The other sampling techniques discussed here all require that auxiliary variable information is available widely across the population. Auxiliary variables that do not have specific numerical values, such as forest type or soil type, can be useful in stratifying the population so that the variation in the variable of interest is less in each stratum than it is in the whole population. When this type of information exists, stratified random sampling might be the most efficient sampling technique.

Where auxiliary variables have specific numerical values and are available, before the ground-based sampling in the inventory starts, from a large number of sampling units spread randomly across the population, sampling with probability proportional to size (PPS sampling) or model-based sampling might be the most efficient sampling technique. PPS sampling would be appropriate when there is a single auxiliary variable only that is correlated positively with the variable of interest. When there are two or more auxiliary variables available, where the relationship between an auxiliary variable and the variable of interest has a complex form or where the correlation between the variable of interest and the auxiliary variable is negative, model-based sampling might be most appropriate. In either case, the higher the degree of association between the auxiliary variable(s) and the variable of interest, the more efficient will be those techniques.

Apart from the sampling technique used, the other determinant of the precision of the inventory (the size of its confidence limit) is the size of the sample. The larger it is, the smaller is the confidence limit likely to be. If a sample has been selected and the confidence limit is larger than is desired, then the sample size must be increased (or a different sampling technique employed) to obtain a smaller limit.

Lastly, it must be stressed that this chapter has only introduced the subject of sampling techniques. Many and much more complex variations exist to deal with the problems encountered in sampling forests; more advanced texts (e.g. Schreuder et al. 1993; Shiver and Borders 1996; Kangas and Maltamo 2006; Gregoire and Valentine 2008) should be consulted to learn more about these. However, the discussion here should give the reader a basic understanding of some important sampling techniques used today in forest inventory.

Chapter 11
Conducting an Inventory

Abstract Numerous practical issues need to be addressed if an inventory is to be carried out to determine successfully the value of some forest characteristic across an entire forest area. This chapter discusses those issues. They include consideration of the objectives of the inventory and the level of certainty required of the results, collation of pre-existing information about the forest, determination of the forest area, the use of square grids as the basis of a random sampling pattern, characteristics of fixed area plots or point samples, problems of plot measurement near forest edges, determination of confidence limits and more complex inventory methods.

11.1 Objectives

Chapters 9 and 10 have established the background to undertaking an inventory of a forested area. The first practical step in doing so is to establish very clearly, with the forest owner, the objectives of the inventory to ensure it will achieve whatever the owner has in mind.

Forest owners differ greatly both in the size and nature of their forest holdings and in the purposes for which they own them. At one end of the scale is the farmer who may own some tens of hectares of forest being used to beautify the farm as an 'environmental' forest or as an investment for retirement. At the other end of the scale are governments or corporations that own large areas of native and/or plantation forests with a myriad of uses, ranging from timber production to water catchment protection, to wilderness or biodiversity conservation (Corona et al. 2011) or to recreation for people.

The scope of the inventory task will differ greatly at these extremes. At the smaller end of the scale, a single person may be able to carry out the inventory. At the bigger end, large teams of both professional and technical staff may be employed permanently to carry out the inventory and analyse and interpret its results. McRoberts et al. (2010) have reviewed the development across the world of inventory practice at the larger scale.

© Springer International Publishing Switzerland 2015
P.W. West, *Tree and Forest Measurement*, 3rd edition,
DOI 10.1007/978-3-319-14708-6_11

Nevertheless, whatever the scale of the inventory, its objectives must be quite clear. They will determine exactly what measurements are to be made and the nature and scale of the sampling necessary to estimate whatever variables are of interest across the whole forest. It may be only wood quantities in the forest that are required. However, for larger and more complex forests, information on many other characteristics of the forest ecosystem may be needed. The methods necessary to measure those different characteristics may differ so greatly that each requires a quite separate inventory. Trees do not move, so they can be easily located and measured on different occasions if necessary. Animals hide, move about and may bite, so sampling techniques necessary to locate and measure them are quite different from those appropriate for trees. If the water resources provided by a forested catchment are to be assessed, sampling may concentrate on stream out-flows and the water holding characteristics of the soils within the catchment. This book concentrates on inventory of tree qualities in forests and particularly on wood and biomass quantities; other specialist texts will need to be consulted if the interest is in inventory of other forest ecosystem characteristics.

Once it has been established clearly with the forest owner what characteristics of the forest are to be estimated, consideration needs to be given to the level of confidence with which the owner wishes to know the answer and, hence, what width of confidence interval will be acceptable. These will determine the amount and type of sampling that will be required.

Thus, an owner concerned principally as to whether or not an endangered plant species occurs within his or her forest area may require only to be 75 % sure that its numbers lie within ±20 % of the estimate of those numbers. A potential forest owner, wishing to value the wood resource in a forest before purchasing it, may wish to be 95 % sure that its quantity lies within ±5 % of the amount estimated by the inventory. The work and costs involved in undertaking the inventory will be greater the more confident the owner wishes to be of the answer.

These statistical concepts are often difficult for lay people to appreciate. Even if the forest owner does not understand them fully, the professional person undertaking the inventory needs to at least infer what the requirements of the owner are and establish what they should be in negotiation with him or her. Only then will it be possible to judge the scope of the inventory task and, hence, how much it will cost. Perhaps even more importantly for the professional, it will allow him or her to defend the quality of the results if a legal dispute should arise with a disgruntled owner about the quantities that were estimated.

11.2 Approach and Methods

It would be possible to consider each and every tree in a forest area as being the population to be sampled in an inventory. Sampling could be done amongst those individuals, and their total number would then be used to convert the results from the sample to the required estimate for the total forest. For any but a very small forest area, this would be impractical. To travel hither and thither over a large forest area to sample individual trees would require far too much time and be unnecessarily arduous. Accordingly, forest inventory is usually carried out by sampling stands scattered throughout the forest area.

Ultimately, an inventory provides an estimate of the stand mean and its confidence limit over the entire forest area for the variable of interest being measured. For example, an inventory might determine that the mean stand stem wood volume over a forest area is 158 m^3/ha, with a 95 % confidence limit of ±17 m^3/ha. To determine the stem wood volume available over the entire forest, and its confidence limit, then simply involves multiplying these values by the area of the forest. Thus, if the entire forest had an area of 8,471 ha, it would be estimated that there was 1.34 million m^3 ($8,471 \times 158/1,000,000$) of stem wood in the whole forest area, with a 95 % confidence limit of ±0.14 million m^3 ($8,471 \times 17/1,000,000$). That is, it would be estimated that we could be 95 % sure that the stem wood volume in the entire forest lay within the range 1.20–1.48 million m^3 (1.34 ± 0.14).

Large forest estates often have a lot of information already collected about them. Maps may exist of forest types, soils, topography or climate variation. There may be airborne or satellite imagery available (Chap. 13). There may be reports available of previous management undertaken, of previous inventories or of ecological research done in the forest. All this information should be consulted and may prove useful in planning the inventory. At least it might provide basic maps of the forest areas being considered. Even better, it might allow stratification of the forest to allow stratified random sampling (Sect. 10.2). At best, it might provide auxiliary variable values widely across the forest estate that could be used for sampling with probability proportional to size (Sect. 10.3) or model-based sampling (Sect. 10.5).

All this information will assist in defining exactly the population that is to be measured. It will help also to make decisions about what sampling technique to use and what the sample size should be; these matters were discussed in Sect. 10.6. Decisions will have to be made too about the measurement techniques to be used for stands included in the sample. Plot measurement (Sect. 8.4.1) might be used, or, increasingly today, point sampling (Sect. 8.4.2) might be preferred; issues involved with measuring plots and point samples are discussed in Sect. 11.5. Ultimately, all these decisions will have to be weighed to reach some compromise as to how to carry out the inventory in the most cost-effective way, given the prior information, equipment and staff resources available.

11.3   Forest Area

It is clearly crucial to the approach described in Sect. 11.2 that the total area of the forest population be determined. If stratified random sampling is to be used, the area of each stratum must be measured also. It is essential to have an unbiased and precise estimate of these areas, since the results of sampling are eventually multiplied by them to determine the final estimate of whatever is being measured over the whole forest. Any error in forest area is immediately transferred to the final result; poor area measurements are often one of the biggest sources of error in an inventory estimate.

For large and complex forests, measurement of area can be a difficult task. It may require the services of professional surveyors, cartographers and **geographic information system** specialists. Particularly difficult is the determination of what actually is, or is not, included in the forest population given the constraints that are imposed by the need for sustainable forest management (e.g. Davis et al. 2001; Furst et al. 2007; Burger 2009; Vanclay 2009; Ryan 2013). For example, logging might not be permitted within a certain distance of streams to avoid siltation or on slopes above a certain steepness. Inventory of timber availability would need to exclude those areas and they would have to be mapped out of the forest area accordingly. In a forest plantation, the trees may not have grown at all on swampy areas or refuge areas of native forest for wildlife may have been left scattered throughout the plantation. Again, these need to be mapped or, if they are not, must be included in the sampling process and wood availability from them recorded as zero when they are encountered as part of the sample actually taken.

There is a variety of more and less sophisticated surveying equipment available these days. It can range from a simple magnetic compass, clinometer and tape to a theodolite and laser distance measuring equipment or to the **global positioning system**. The precision of the survey would be expected to increase as more sophisticated equipment is used. Nevertheless, the principles of the conduct of a survey remain the same, no matter what equipment is used; these are outlined in Chap. 12.

After the survey has been completed, the ready availability today of geographic information systems has made much easier the process of producing finished maps of complex forest areas. These systems are used universally today by organisations routinely involved with forest inventory. Description of their use is outside the scope of this book.

11.4 Sampling Pattern

All sampling techniques require that the sample be selected objectively through some random selection process (Sect. 9.2). There is no inherent reason why the locations of the sampling units should not be selected simply at random, that is, scattered here and there right across the population. However, it can be rather tedious practically for measurement crews to navigate to random locations. As well, by chance, a set of random locations does not always provide the more or less even spread of locations across the population that leaves the sampler feeling comfortable that the entire population has been considered adequately in the sampling process.

Given these issues, it is common in forest inventory to position a square or rectangular grid over the entire forest population area as the basis of the sampling scheme. The grid need not actually be square or rectangular; any geometrically regular shape would do. However, square or rectangular grids are usually easiest to deal with in practice. The intersection points of the grid constitute a set of locations across the population at which samples might be taken.

The important thing about setting up such a grid is that its starting point should be chosen randomly. If this is done, each and every point in the entire population has the same chance of being included as an intersection point on the grid. That is, the grid defines a set of points that constitute a simple random sample of locations across the population (Sect. 10.1). All the intersection points, or a randomly selected subset of them, may than be used as a simple random sample. Navigation by measurement crews on such a grid is generally quite easy and the sampler can feel that the grid covers thoroughly and evenly the entire area of the population.

The decision about the size of the grid squares or rectangles is left up to the person conducting the inventory. Usually a size will be chosen to give an appropriate number of sample points at which auxiliary variables and/or the variable of interest to the inventory will be measured. It is common in modern forest inventory that aircraft are used to measure auxiliary variables by flying along a grid of strip lines spaced regularly over the forest area. This provides a large set of auxiliary variable values along these strip lines. Field sampling of the more difficult to measure variable of interest would then take place at a randomly selected subset of sample points along those strip lines, the random selection of points being done in a fashion appropriate to the sampling technique being used (Chap. 10).

A slight variation of this grid sampling process is known as systematic sampling. It is a system used often in America where many foresters make their living assessing timber availability from privately owned tracts of forests; American foresters term this 'timber cruising' or 'cruising'.

Systematic sampling also involves laying out a randomly oriented grid over the forest area, but with a grid size such that the number of intersection points equals the sample size required. Each and every intersection point is then sampled. Because there is then no selection of which points are to be included in the sample, it is a sampling technique appropriate only to simple random or stratified random sampling.

Systematic sampling is straightforward to apply in the field. The sampler starts at a randomly chosen point, somewhere near a corner of the forest tract, and walks along straight lines directly through the forest measuring a plot or taking a point sample at regular intervals as determined by the grid spacing.

Foresters who do this type of work regularly know from experience what an appropriate size is for the grid for the particular forest type within which they are working. The sample size is then determined directly by the area of the forest. The forest area itself can be estimated by multiplying the sample size by the area of the grid squares, so avoiding the need for a separate survey to determine the forest area. The sampler might also be able to identify stratum boundaries as he or she walks through the forest, thus effectively mapping the forest as he or she goes. Formally speaking, there are some mathematical statistical considerations to be taken into account when considering the use of systematic sampling. Valentine et al. (2009) have reviewed these issues.

11.5 Stand Measurement

Stands are usually measured in forest inventory using plots or point samples. In doing so, a number of issues need to borne in mind. Some of these are important for both plot measurement and for point sampling, and others only for one or the other.

11.5.1 Shape

Plots are usually square, rectangular or circular. In principle, there seems no reason in forestry to prefer one shape over the other.

There are difficulties involved with all shapes. Circular plots may be easier to establish, since their circumference can be readily identified by running a string from the centre out to the required radius. However, because the boundary is curved, it is more difficult to decide which trees are inside or outside the plot. Square and rectangular plots can be more time consuming to lay out because they have to be established carefully with right angles at the corners. Their boundary length is also greater than that of a circular plot of the same area, so more decisions have to be made as to which trees are actually in or out of the plot; on the other hand, their straight edges can make these decisions easier. As mentioned in Sect. 8.4.4, it is usually assumed that a tree is in the plot if the centre of its stem lies within the plot boundary.

Point sampling inherently involves measuring trees from circular areas (Fig. 8.1), but care needs to be taken when deciding if borderline trees should be included or excluded (Sect. 8.4.3).

11.5.2 Positioning

Whichever plot shape is preferred, additional care needs to be taken in plot positioning when the trees are positioned with some regular arrangement, such as in a plantation. Because of the regular spacing, it is possible to position plots of the same area but that contain different numbers of trees. However, it is usually quite straightforward to find a consistent way of positioning plot corners or the plot centre and choosing the plot area so that the stocking density of the trees in the plot is similar to that of the general stocking density of the plantation.

In plot or point sampling, care must be taken where forests have some regular trends in tree size imposed on them by physical characteristics of the land or management practices. For example, where logging debris has been heaped into long rows (often called windrows) and burnt, trees in the regenerated or replanted forest may grow better on the windrows because of the soil sterilisation and nutrient release resulting from the fire. If samples were taken at spacings consistent with the windrow spacing, windrows might be sampled either with higher or lower frequency than should be the case.

Trees often grow more poorly on ridges in the landscape, because water availability from the soil is often less on ridges than in the down-slope valleys. If grid lines for a systematic sample happened to be orientated so they were parallel to ridges, areas of poorer or better forest might be sampled more or less frequently than they should be.

Whatever these potential sources of bias in sampling, care needs to be taken to orient the sampling unit grid to avoid them. Sometimes long and narrow rectangular plots, oriented with their long axes perpendicular to the regularity, are used to minimise these problems.

It is common also to come across treeless gaps in the forest. They are part of the forest population and must not be ignored. They can be mapped out of the population or, if not, must be included in the sampling process and recorded as having a zero value of whatever is being measured in the inventory.

11.5.3 Size

Deciding what the area of plots should be presents several problems. In Sect. 10.6, the size of the sample to be taken in an inventory was considered, that is, the number of stands to be measured. However, ultimately it is the intensity of sampling that is rather more important than just the number of sampling units. The intensity is defined as the total area sampled divided by the total forest area. It depends on the number of plots included in the sample, their areas and the total forest area. A certain intensity of sampling will be necessary to achieve any desired precision of the final estimates (i.e. size of confidence limit) sought in the inventory.

It is cheaper generally to measure fewer, larger plots than many smaller plots; the time and cost involved in moving measuring crews from plot to plot is usually much greater than taking more measurements at any one plot. However, if the sample size (i.e. the number of plots sampled) is too small, the sampling may not cover adequately the range of variation that occurs across the forest. Having said that, in forestry practice plot sizes are generally chosen to be within the range 0.01–0.1 ha.

Because plot area is unknown with point sampling, sampling intensity cannot be determined. The smaller the basal area factor used for the point sample, the larger will be the number of trees measured around any point. Again, a balance will need to be drawn between the number of plots sampled and the number of trees measured in each.

If a lot of prior information is available about the forest to be inventoried, it may be possible to undertake computations to determine the optimum balance of plot size, sample size and sampling technique to achieve the most cost-efficient inventory possible. These techniques are outside the scope of this book.

11.5.4 Edge Plots

In any inventory, it is inevitable that some sample locations will be positioned near the forest edge so that a plot established at that point would extend partially outside the forested area or trees would be missing from a point sample as the observer looked beyond the edge. Because they have fewer surrounding trees to compete with them, trees along the forest edge generally grow differently from those well within the forest; they are often larger with bigger branches extending out into the open space beyond the edge. Such trees should be included in the sampling or else some bias in the final results would be expected. Thus, it would be quite inappropriate simply to move a plot or point sample position further inside the forest to avoid an edge overlap. Šálek et al. (2013) gave an example showing the magnitude of this effect of the forest edge on tree development.

A number of methods have been developed to ensure that sampling is conducted properly near forest edges (Schreuder et al. 1993; Avery and Burkhart 2002; Iles 2003; Ducey et al. 2004; Kangas 2006; van Laar and Akça 2007; Gregoire and Valentine 2008). Failure to use such methods will introduce bias into the estimates obtained ultimately from the inventory. Two methods, the mirage technique (Schmid-Haas 1969) and the walk-through technique (Ducey et al. 2004) have become widely used (Iles 2003, Chap. 14). The principles of these two methods will be described here, together with the walk-through and fro technique, a new method that has been developed recently (Flewelling and Strunk 2013).

If fixed area plot sampling is being used and part of a plot falls beyond the forest edge, the mirage technique involves mirroring the segment of the plot lying beyond the edge back inside the forest. Trees falling within the mirrored section are measured again and included twice in the plot measurements. The plot retains its full area. If point sampling is being used, the position of the point is mirrored from the edge into the open space beyond the edge and the point sample made from there as well as from the original point, so also measuring some trees twice. The technique is illustrated in Fig. 11.1. Intuitively, it may seem that measuring the same trees twice is likely to introduce bias into the results, but formal statistical analysis has shown that failing to do so will certainly lead to bias; this matter is discussed in the literature referred to in the previous paragraph.

In the walk-through technique, the sampler walks from the plot centre or point sample point towards any tree within the plot or included in the point sample and then continues the same distance beyond the tree in the same direction. If the sampler is then positioned beyond the forest edge, the tree is included twice in the sample. The technique is illustrated in Fig. 11.2.

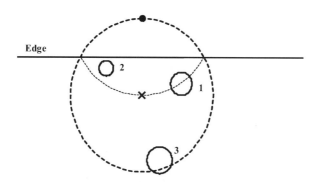

Fig. 11.1 Application of the mirage technique when fixed area circular plots are being used for sampling. The view is from above. The plot is centred at X and its boundary is shown (*dashed line*). The stem cross sections are shown of three trees, numbered 1–3, that are located within the plot. Part of the plot extends beyond the forest edge making it an edge plot. The plot centre has been reflected through the edge to the position ●. The boundary of the reflection (or mirage) of the plot from that point back within the forest is shown (*dotted line*). The centre of tree 2 is positioned within that mirage plot and would be the only tree of the three that would be included twice in the sample. If point sampling was being used instead and a point sample was taken at X with the three trees being counted in the sample, the sampler would then move to the position ● and would take a new point sample from there. Any trees counted in that new sample would be included as part of the original count; this may include any or all of the originally counted trees and may include other trees not included in the original count

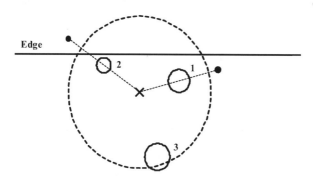

Fig. 11.2 Application of the walk-through technique when fixed area circular plots are being used for sampling. The situation is the same generally as in Fig. 11.1. The sampler 'walks through' from the plot centre (X) towards a tree and continues, on the same line, the same distance beyond it to the point ●. If the sampler is then beyond the forest edge, that tree is included twice in the sample. In this case, only tree 2 would be included twice. The procedure is exactly the same if point sampling was being used instead

The walk-through and fro technique is rather more complex to apply in practice. It requires consideration of what is known as the 'inclusion zone' for any tree that is located within a fixed area plot or included in a point sample. This zone is an area about the tree within which the centre point of the plot or the point sample point must lie for the tree to have been included in the plot or point sample in the first place. For circular fixed area plots, the inclusion zone of a tree is a circular area centred about the tree with a radius equal to the plot radius. For a point sample, it is a circular area centred about the tree with a radius (r, m) equal to $r = D/(2\beta^{1/2})$, where D is the diameter at breast height over bark (cm) of the tree and β (m²/ha) is the basal area factor being used for the point sampling; note that this is the same formula as used in Sect. 8.4.3 in deciding if a tree is to be included or not in a point sample.

Having established these inclusion zones in the field at the time of sampling, the sampler then considers each tree in turn. From the plot centre or sample point, the sampler walks towards the tree and then continues beyond it in the same direction measuring the distance from the tree to either the forest edge or the inclusion zone edge, whichever is closest. This gives what is termed the 'walk-through' distance (t). The sampler then returns to the tree and walks towards the plot centre or sample point and continuing on to measure the distance from the tree to either the forest edge or the inclusion zone edge, whichever is closest. This distance (f) is termed the 'walk-fro' distance. The process is illustrated in Fig. 11.3.

Once this process has been completed, the tree is assigned a 'weight', w, calculated as

$$w = 2r^2/\left(t^2 + f^2\right). \tag{11.1}$$

Any tree for which its inclusion zone does not extend beyond the forest edge, such as tree 3 in Fig. 11.3, would be assigned the value 1 for its weight by this process. Trees with inclusion zones that extend outside the forest edge, such as tree 2, may be assigned a value for its weight that exceeds 1.

Suppose that in fixed area plot of area A (ha) or in a point sample there were n trees in the plot or counted in the point sample. Suppose the diameter at breast height over bark of the ith of those trees was D_i (cm), its stem volume (or tree biomass) was V_i (m³ or tonne) and the weight assigned to it through the walk-through and fro process was w_i. For a fixed area plot, its stocking density (S, stems/ha) would then be determined as

$$S = \left(\sum_{i=1\ldots n} w_i\right)/A, \tag{11.2}$$

its stand basal area (G, m²/ha) as

$$G = (\pi/40,000)\left(\sum_{i=1\ldots n} w_i D_i^2\right)/A \tag{11.3}$$

and its stand volume (or biomass) (V, m³/ha or tonne/ha) as

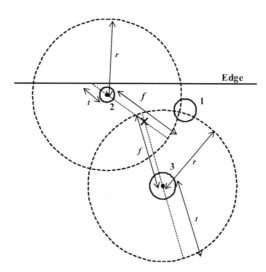

Fig. 11.3 Application of the walk-through and fro technique when fixed area circular plots are being used for sampling. The situation is similar to that of Fig. 11.1, although the boundary of the plot (that is centred at X) is not shown here. The use of the walk-through and fro technique is described for trees 2 and 3; it would be applied to any other tree (such as tree 1) that was included in the plot area. The circular plots (*dashed line*) drawn here represent the inclusion zone of each tree; they are centred on the tree concerned and are equal in radius (*r*) to the circular plot used for sampling. The lines walked by the sampler in applying the technique are shown (*dotted line*) and the 'walk-through' (*t*) and 'walk-fro' (*f*) distances measured are shown for both trees. The only difference from this process for point sampling would be that the radius of the inclusion zone of each tree included in the original tree count would be determined separately as described in the text

$$V = \left(\sum_{i=1...n} w_i V_i \right) / A. \tag{11.4}$$

If point sampling had been used with a basal area factor of β (m^2/ha), stocking density would be determined as

$$S = \beta(40,000/\pi) \left(\sum_{i=1...n} w_i/D_i^2 \right), \tag{11.5}$$

stand basal area as

$$G = \beta \left(\sum_{i=1...n} w_i \right) \tag{11.6}$$

and stand volume (or biomass) as

$$S = \beta(40,000/\pi) \left(\sum_{i=1...n} w_i V_i/D_i^2 \right). \tag{11.7}$$

The description above of these three edge plot measurement techniques is somewhat idealised. Firstly, in Figs. 11.1, 11.2 and 11.3 it has been assumed that the forest edge is straight. That is certainly not always the case. Both the mirage and walk-through techniques can be far more difficult to apply if the edge is convoluted or the plot or sample point is positioned near a corner of the forest (Iles 2003; Gregoire and Valentine 2008). Secondly, the walk-through method may require the sampler to walk beyond the forest edge. If the edge happens to be adjacent to a cliff or a lake, it may be impossible to do so. Thirdly, both the mirage and walk-through methods may introduce some bias into the stand estimate under some circumstances (Iles 2003; Gregoire and Valentine 2008). Fourthly, it has been assumed that a circular fixed area plot was being used. If rectangular plots are used, it can be rather more complex to determine how they should be reflected about the forest edge (Ducey et al. 2004). Finally, if rectangular plots are used, determination of the inclusion zone for each tree is rather more complex (Gregoire and Valentine 2008). Having said that, it is the walk-through and fro method that deals most adequately with these various problems. It is an unbiased technique under all circumstances, does not require the sampler to walk beyond the forest edge and can be applied readily no matter how convoluted the shape of forest edge. However, it is more complex to apply than the other techniques and is newly developed so it has had little application in practice to date.

For many forest inventories over broad-scale forest areas, the majority of the samples will be taken at points well within the forest, and these edge techniques will only need to be used occasionally. However for complex forest areas that may be scattered in relatively small patches across the landscape, that may contain many treeless gaps scattered through them or that may have watercourses meandering through them, there may be very extensive lengths of edges. Under those circumstances, it may be necessary to use these techniques frequently.

When edge techniques are required for a relatively high proportion of samples in an inventory, it can be expected that the precision will be reduced of the estimates made of the population mean or total of whatever forest characteristic is being measured. This issue was explored by West (2014b) who found that this was indeed so, but the loss of precision seemed to be sufficiently small as to be of little practical consequence for many inventories, even when a very high proportion of the samples required the use of the edge techniques. Further, he concluded that none of the techniques was superior to the others in this respect. He suggested that any or all of the three techniques could be used in the one inventory depending on the circumstances of any particular sample. Thus, the easier to apply mirage or walk-through techniques might be used near an edge that did not have a complex shape, whilst the more difficult to apply walk-through and fro technique might be reserved for use with edges of a complex shape.

11.6   Confidence Limits

For the different sampling methods described in Chaps. 9 and 10, methods have been given to determine the confidence limit about the estimate of the population mean or total derived from the sample. This uncertainty in the estimate is a consequence of measuring only a sample from the population rather than measuring the entire population. In mathematical statistics this uncertainty is known as the 'sampling error'. Here, the term 'error' is being used in its mathematical statistical sense; it does not mean a 'mistake', but rather that the method (in this case sampling) used to get the estimate of the population mean or total has led to some uncertainty in the value obtained.

It is important to realise that sampling error is not the only source of uncertainty. In all the discussion on inventory to here, it has been assumed that the variable of interest being measured in a sampling unit can be measured simply and directly so that it is unbiased and extremely precise. That is not always so. There are several ways in which uncertainty may enter the values that we obtain when we measure or estimate some characteristic of a sampling unit (Cunia 1965), uncertainty that may be termed 'measurement error'. Firstly, genuine mistakes may occur. For example, mis-identification of a species in a complex forest is possible, even when experienced botanists are involved in the inventory. In any large inventory, some such mistakes are perhaps inevitable even when every effort is made to minimise their occurrence.

A second source of measurement error arises from imperfections in measuring equipment and its users; the same answer will not always be obtained by different people measuring the same thing with the same piece of equipment. An example where such a problem might easily occur would be measurement of an upper stem diameter of a tree using an instrument such as a Relaskop (Sect. 5.3.4). Such instruments require care and skill in their use, but even then are certainly less precise than measuring stem diameters with a girth tape (Sect. 3.3).

Thirdly, we often use estimation tools such as volume or biomass functions to estimate the size of individual trees from measurements of simple characteristics such as tree stem diameter and tree height (Chaps. 6 and 7). Such functions predict an average value for trees of any particular diameter and height and there is inevitably some uncertainty about the estimate they give for any particular tree. Wutzler et al. (2008) found that when their biomass estimation function for individual trees of European beech (*Fagus sylvatica*) across central Europe was used to predict stand above-ground biomass (by summing biomass estimates for individual trees in the stand), the estimates of stand biomass were highly likely to lie within just a few per cent of their true values. This is a comfortingly high level of precision, given the extent to which biomass estimation functions are being used around the world. Berger et al. (2014) gave an interesting example of the uncertainty in estimates of tree stem volumes in forest inventory in Austria that arose both from the volume function used and the need to take upper stem diameter measurements.

In statistical terms, these uncertainties in measurements or estimates of sampling unit sizes ultimately increase the estimate of the variance of the population mean, $V(\overline{Y})$ (Chaps. 9 and 10). In turn, this will increase the size of the confidence limit determined for population estimates.

Fortunately, it may sometimes be possible to simply ignore the problem. The uncertainty associated with estimates from a well-constructed volume or biomass function is often so small as to be negligible in comparison to the size of the variation between the stand volumes or biomasses between the different sampling units in the population. McRoberts and Westfall (2014) gave a good example where it was certainly reasonable to ignore the problem. They described an inventory of an area of native forests, consisting of a mixture of 38 hardwood and softwood tree species, covering some tens of thousands of hectares in Minnesota in northern USA. In a simple random sample of 2,178 plots, each 0.067 ha in size, the diameters at breast height and total heights of all trees with a diameter of at least 12.7 cm were measured. Functions to predict stem volume from diameter and height were available for the various species, based on other works in the region where stem volumes had been measured directly for over 2,000 trees. When the error in the volume functions was ignored, McRoberts and Westfall estimated that the average stand volume of the forests was 90.3 m³/ha with a 95 % confidence limit of ±3.7 % of this average (see the last two columns of the first row of their Table 2). They tested various methods of developing their volume functions, and, when they incorporated the error from the crudest method they tried, the average stand volume was estimated as 90.5 m³/ha with a 95 % confidence limit of ±4.3 % of this average (see the last two columns of the last row of their Table 2). Thus, even with their crudest volume function, there was only a small increase in the size of the confidence limit; when they used more sophisticated methods to derive their volume functions, there was virtually no effect on the confidence limit of incorporating the volume function uncertainty.

This book is not the place to discuss in detail the mathematical statistical procedures necessary to incorporate measurement error when it is too large to ignore. Particularly as methods are developed to measure forests using remote sensing from the air or satellites (Chap. 13), it is likely that measurement error will become increasingly important in determining the uncertainty of estimates from inventory. Studies on the topic and development of mathematical and statistical methods to deal with it are continuing (Gertner 1990; Parresol 1999; Zianis 2008; Yanai et al. 2010; Djomo et al. 2011; Gregoire et al. 2011; Ståhl et al. 2011; Cohen et al. 2013; Holdaway et al. 2014; McRoberts and Westfall 2014).

Of recent times there has been considerable interest in mapping forest stand volume or biomass over large areas and even over whole countries. These inventories are generally based on auxiliary information obtained from satellite or aerial imagery (Chap. 13). They are sometimes coupled with a ground-based sample of plots, where the variable of interest has been measured or estimated directly, or with a model system that predicts the variable of interest from the auxiliary data. It is perhaps of some concern that often the uncertainty (i.e. the confidence limit) about the estimates made at any point across the landscape is not determined formally in this work. This may lead users of these maps to have more confidence in their results than is justified. Examples of this type of inventory can be found in Latta et al. (2009), Nolé et al. (2009), Lasserre et al. (2011), Du et al. (2014) and Muinonen et al. (2014). Mitchard et al. (2014) and Ometto et al. (2014) have examined a number of estimates that have been made of forest biomass across the Amazon Basin and shown clearly that there is considerable uncertainty amongst the various results. Research is now attempting to address this problem (Hill et al. 2014).

11.7 More Advanced Inventory

Chapters 9–11 have provided only a basic introduction to the way in which forest inventory is carried out today. However, for relatively simple forests, for areas that are not very large and for variables of interest that are not too difficult to measure or estimate, the techniques discussed in this book should be adequate for many inventory purposes.

Much more sophisticated sampling techniques and methods of conducting the inventory are used from time to time in different forest types in different parts of the world and to assess different forest characteristics. These techniques may involve several stages of sampling. Perhaps satellite images might be used to obtain some information from a sample of the forest. Air photos, being of a smaller scale, might be able to provide more detailed measurements on a smaller subsample. A still smaller ground sample might then produce highly detailed measurements of the variable of interest ultimately to be determined for the forest. The earlier stages may have then provided auxiliary variable values that may be related to the variable of interest. The larger and more complex the forest concerned, the more it will be necessary to adopt these advanced techniques. Tomppo et al. (2014) have presented a good example of the methods involved in planning an inventory over whole country, in their case an inventory of the forests of Tanzania.

Chapter 12
The Plane Survey

Abstract For inventory and general forest management purposes, it is essential to have available a map of any forested area of interest. This chapter describes how a survey of the boundary of a forested area should be conducted to produce a map drawn on a flat (plane) surface. The principles of the global positioning system are described also, a system that has use for navigation around the forest as well as for the conduct of surveys.

12.1 Mapping

For most forest inventories, the area of the forest must be determined (Sects. 11.2 and 11.3). Generally, this involves surveying the forest boundary, drawing a map from the survey and determining the forest area from the map.

Mapping is properly the realm of the professional surveyor and cartographer. Much more information may be included on maps than just the boundaries of areas of interest. However, anyone working in forestry needs to understand the principles at least of how a survey of a forest area is done, how a map of its boundaries is drawn from it and how the area of the forest is calculated.

Highly sophisticated instruments, including precision theodolites, laser distance measuring equipment and global positioning system receivers, are the tools of trade necessary to conduct more rigorous surveys. The global positioning system has become a generally useful tool for many purposes in forestry and is discussed in Sect. 12.6. Computer-based geographic information systems are available readily to draw the maps from surveys and calculate the areas of mapped regions; they are used by most forestry organisations today. However, using even simple instruments, a good-quality magnetic compass, a clinometer and a measuring tape, any forester should be able to conduct a reasonable survey of a forest area and draw a useful, basic map of it.

The main theme of this chapter is to establish the principles of a plane survey of a modestly sized parcel of land. It is termed a plane survey because the objective will be to draw a map of the boundaries of the surveyed area on a flat piece of paper, that is, on a plane. This means that wherever the land is sloping, the distances measured

© Springer International Publishing Switzerland 2015
P.W. West, *Tree and Forest Measurement*, 3rd edition,
DOI 10.1007/978-3-319-14708-6_12

Fig. 12.1 Map of a section of land that is to be surveyed. The total distance around its perimeter is 752.5 m and the area enclosed is 1.43 ha

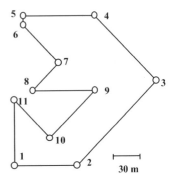

along the slopes will all have to be reduced to horizontal distances; this is the way most maps are presented. Where it is wished to show the topography of the land on a plane map, it is usually presented as contour lines, each contour connecting points on the map that are at the same altitude. However, we will not consider here how contour maps are drawn.

12.2   Survey Example

Figure 12.1 shows a simple, plane map of a rather oddly shaped piece of land that encloses an area of exactly 1.43 ha. This area will be used as an example to illustrate how a plane survey of it could be carried out and what computations need to be done to convert the survey measurements to a map that looks like Fig. 12.1.

Figure 12.1 was drawn with X- and Y-axes that had their origin exactly at Point 1 and with the Y-axis running vertically along the line that joins Point 1 to Point 11. The exact X- and Y-coordinates of each of the 11 points around the boundary are shown in the last two columns of Table 12.1. This chapter illustrates how the measurements taken in a survey of this area are converted to a set of X- and Y-coordinates so that a map of it could be drawn easily.

12.3   Conducting the Survey

In principle, a plane survey is carried out by starting at any arbitrarily chosen corner point around the area to be surveyed. The surveyor then moves progressively from corner point to corner point around the survey area. The distance between each pair of corner points, the slope of the land between them and the angle by which the direction of travel changes at each point are measured. Usually at least two people are required to conduct a survey, both to move the equipment and take the required measurements.

The distances between points can be measured with a tape (many years ago a light chain that could not stretch with repeated use was used to measure distances and gave us the name of the old imperial unit 'link' for distance measurement). For more precise distance measurements, laser measuring devices are available today and avoid the need to have to drag the tape along the ground between measurement points. Using a global positioning system (Sect. 12.6) receiver also avoids the need to use a tape.

For more precise surveys, a theodolite may be used to measure the angle of direction change at each point. In the example in Fig. 12.1, this would be done at Point 2, say, as follows. The theodolite is mounted on its tripod immediately over Point 2. The surveyor looks back to Point 1 through the telescope of the theodolite and then rotates the instrument to view Point 3. The theodolite scale then gives the angle through which the instrument had been rotated and this is the angle of direction change required at Point 2. It will be assumed in this book that any such angles are measured clockwise from the line looking back to the preceding point; it does not matter if angles are measured anticlockwise, as long as the surveyor is consistent throughout the survey.

Angles can be measured also with a compass. In this case, standing at Point 2, the surveyor measures the bearing from north to Point 3 to give the required direction change. Some theodolites have an inbuilt compass and can be used to measure these bearings. Good-quality, hand-held compasses can be used also, but these would not be expected to be as precise as a theodolite. A global positioning system receiver can also be used to measure bearings. For many forestry purposes, where a highly precise map is not required, a hand-held compass serves adequately. When using a compass, the user must ensure that no metal objects, such as metal fence posts, are nearby because they may influence the compass reading.

Theodolites generally allow vertical as well as horizontal angle measurements. Thus, standing at Point 2, the surveyor measures the vertical angle up or down to Point 3. This gives the required slope angle to the next point. To ensure the measurement is correct, the surveyor must sight to a point on a staff, mounted at Point 3, the same height above ground as that of the theodolite eyepiece. Hand-held clinometers can be used to obtain slope angles also. These are less precise than a theodolite but serve adequately if a high degree of precision is not required of the survey.

If the slope angle changes appreciably (say by more than 2–3°) along different parts of the line between any two points, the surveyor needs to establish new survey points where the slope changes; since the distance between points is measured as the distance along the slope (the tape is laid along the sloping ground), it is obviously important that the slope between any two measurement points should not vary greatly at any point between them. If there is a gully between two survey points, it is necessary to establish a new point at the bottom of the gully to deal with the different slopes on each side of it. Use of a global positioning receiver avoids the need to determine slope angles because it provides information to allow automatic determination of the horizontal distance between survey points.

Table 12.1 Survey data and results of computations to find corresponding map coordinates

Survey point	Direction change at point (α_i) (degrees)	Angle line to next point makes with Y-axis (β_i) (degrees)	Slope angle to next point (γ_i) (degrees)	Slope distance to next point (s_i) (m)	Horizontal distance to next point (h_i) (m)	X-coordinate determined from survey data (x_i) (m)	Y-coordinate determined from survey data (y_i) (m)	Actual X-coordinate (m)	Actual Y-coordinate (m)
1	**90**	90	**−8**	**61.5**	60.9	0	0	0	0
2	**136**	46	**0**	**128.5**	128.5	60.9	0.0	70	70
3	**89**	315	**8**	**100.0**	99.0	153.3	89.3	160	90
4	**134**	269	**3**	**70.8**	70.7	83.3	159.3	90	160
5	**89**	178	**−8**	**18.2**	18.0	12.6	158.1	10	160
6	**136**	134	**−2**	**55.0**	55.0	13.3	140.0	10	150
7	**269**	223	**−3**	**41.0**	40.9	52.8	101.9	50	110
8	**45**	88	**3**	**73.9**	73.8	24.9	71.9	20	80
9	**316**	224	**1**	**68.5**	68.5	98.6	74.5	90	80
10	**271**	315	**3**	**60.0**	59.9	51.0	25.2	40	30
11	**46**	181	**−1**	**62.0**	62.0	8.7	67.6	0	70
1						7.6	5.6	0	0

Values in **boldface** are measurements taken in the survey. All other values were calculated from them. The coordinates in the seventh and eight columns were determined from the survey measurements and are plotted in Fig. 12.2. The coordinates in the last two columns are the actual coordinates, as in Fig. 12.1

Table 12.1 lists the set of measurements that a survey team might make in conducting a survey around the area in the example in Fig. 12.1. It was assumed in this case that a theodolite without a compass was used to measure the change of direction at each point, so the angle measured was taken by looking back to the preceding point and turning the theodolite round to view the next point. The actual measurements taken are shown in boldface type in the table. All the other values shown were computed from them as will be discussed below.

12.4   Calculating the Survey Results

At the start of the survey, an arbitrary decision needs to be made as to the coordinate system in which the final map is to be graphed. For convenience, the starting point of the survey (Point 1 in the example) is usually chosen to be the origin, that is, to have X- and Y-coordinate values of 0,0. It is convenient also to assume that the Y-axis runs along the line joining the first point to the last one (the line joining Points 1 and 11 in the example), with Y-coordinate values positive and increasing towards the last point. The X-axis is then perpendicular to the Y-axis through the origin.

Given these assumptions about the axes, it is possible to take the survey data, point by point in the order in which they were measured, and determine X- and Y-coordinate values for each survey point. The final map of the survey boundary can then be drawn on graph paper (or by using any of the numerous computer software packages now available to do this) using the calculated coordinate values.

For the ith survey point ($i = 1 \ldots 11$ in the example), let the angle of the direction change at that point be α_i and the angle (clockwise from the positive direction of the Y-axis) that the line to the next survey point makes with the Y-axis be β_i. Let the angle of the slope of the ground to the next point be γ_i (it is positive for an upslope and negative for a downslope), the distance measured along the slope to the next point be s_i, the corresponding horizontal distance be h_i and the X- and Y-coordinate values calculated for the point be x_i and y_i. The measurements taken in the survey provide values for α_i, γ_i and s_i for each of the survey points. Values for all the other variables must be calculated from them.

The computations begin by considering the first survey point. Because of the way it was assumed the axes were positioned, with their origin at that point, values of the various variables for that point are determined automatically. So $\beta_1 = \alpha_1$ ($= 90°$ in the example) and $x_1 = y_1 = 0$. These values can be entered immediately in the results table and are shown in the first row, for Point 1, of Table 12.1. The horizontal distance, corresponding to the slope distance, to the next point is calculated as

$$h_i = s_i \cos (\gamma_i) \tag{12.1}$$

For the example, this gives $h_1 = 61.5 \times \cos(-8) = 61.5 \times 0.99027 = 60.9$ m, the result shown for Point 1 in Table 12.1.

The results for the remaining survey points are now considered in the order in which they were surveyed. For each, the horizontal distance to the next point is calculated using (12.1). The remaining results each depend on the results calculated for the preceding survey point. The angle that the line from the ith point to the next survey point [the $(i+1)$th point] makes with the Y-axis, β_i ($i=2\ldots11$ in the example), is calculated as

$$\beta_i = \beta_{i-1} + \alpha_i - 180, \tag{12.2}$$

where all angles are in degrees (if other angular units are being used, the $180°$ in (12.2) would have to be replaced by the corresponding value for whatever angular units are being used). The result from (12.2) is often negative (representing an angle measured anticlockwise from the Y-axis). It can be left in that form because the subsequent trigonometric calculations give the same answers whether the angle is positive or negative. However, if desired, a negative answer can be converted to the same angle, expressed as a positive angle measured clockwise from the Y-axis, simply by adding $360°$ to the result; this conversion has been done wherever appropriate in calculating the results in Table 12.1. In the example, (12.2) gives $\beta_2 = 90 + 136 - 180 = 46°$. This is the result shown in the second row of the table for survey Point 2.

The X- and Y-coordinates of the ith point are calculated as

$$x_i = h_{i-1} \sin(\beta_{i-1}) + x_{i-1} \tag{12.3a}$$

and

$$y_i = h_{i-1} \cos(\beta_{i-1}) + y_{i-1} \tag{12.3b}$$

So, for the example, $x_2 = 60.9 \times \sin(90) + 0 = 60.9 \times 1 + 0 = 60.9$ m and $y_2 = 60.9 \times \cos(90) + 0 = 60.9 \times 0 + 0 = 0$ m, results shown for survey Point 2 in the second row of the table.

These computations are repeated for all the remaining survey points and are shown in the table. In addition, the computations are repeated one last time, after completing the results for the last survey point, Point 11 in the example. This gives a new pair of coordinates for the original starting point, Point 1 in the example. The resulting X- and Y-coordinates, 7.6 and 5.6 m, respectively, are shown in the last row of the table. They are not the same values 0 and 0 m that we know are the actual coordinates of Point 1 since it was chosen as the starting point. The difference is inevitable because the survey can never be carried out perfectly. The angles and distances will never be measured exactly because of limitations in the measuring devices and the limitations of the people making the measurements.

Figure 12.2 shows the final plotted survey, using the coordinate values calculated from the survey data. It can be compared with the original in Fig. 12.1. The deviation of the final calculated position of survey Point 1 from its original position is obvious.

Fig. 12.2 Plotted result
after conducting a survey of
the parcel of land depicted
in Fig. 12.1

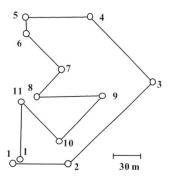

The distance between the final, calculated position of the starting point and its known position is known as the closing error of the survey. The distance can be calculated using the general function used to calculate the length of the straight line, t, that joins any two points of which the X- and Y-coordinates are known, (x_a, y_a) and (x_b, y_b), where

$$t = \sqrt{\left[(x_a - x_b)^2 + (y_a - y_b)^2\right]} \qquad (12.4)$$

For the example, $t = \sqrt{[(7.6 - 0)^2 + (5.6 - 0)^2]} = 9.4$ m. That is to say, at the end of the example survey, the calculated position of the starting point of the survey was 9.4 m away from where it should have been due to the errors made in taking the measurements.

In conducting the survey, a total horizontal distance of 737.2 m was travelled around the perimeter of the area being surveyed (the sum of the values h_i in Table 12.1). That is to say, after travelling 737.2 m, we ended up 9.4 m away from where we should have been. So, in conducting the survey, we were 1 m away from where we should have been for every 737.2/9.4 = 78.4 m travelled. This quantity, the traverse distance divided by the closing error distance, is known as the accuracy of the survey and is used generally by surveyors to judge the quality of their work.

An accuracy of 1 m in 78.4 m would be considered inadequate by most professional surveyors who usually pride themselves on achieving much higher accuracies. However, it depends entirely on the purpose for which the survey is being done as to what accuracy is required. In the example, the missing 9.4 m of land represented by the closing error might lead to a bitter dispute between neighbours as to where the boundaries between their properties lay; for legal purposes, a surveyor would have to be more accurate than that. By contrast, for a forester making a rapid assessment of the area of a plantation, it might be quite adequate.

In the example survey, it was assumed that a theodolite was being used and the angle of direction change at each survey point was measured. If a compass (or global positioning system receiver) was being used, the bearing from north of

Fig. 12.3 Map of the
section of land shown in
Fig. 12.1, divided into
triangles for area
determination

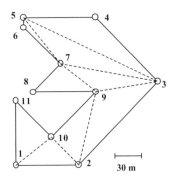

the next survey point would be measured instead. Under these circumstances, it would be assumed usually that the Y-axis of the coordinate grid would be directed to magnetic north rather than along the line joining the first and the last survey points as assumed in the example. The bearings from north would then be direct measures of the β_i in Table 12.1, and no values of the α_i would be recorded in the survey. In calculating the coordinates of the survey points, there would then be no need to use (12.2).

12.5   Area of a Surveyed Region

Once a survey is complete, it is often desired to calculate the area enclosed by the survey. A simple way to do this is to divide the area into triangular subsections, calculate the area of each triangle and sum them to give the total area. Figure 12.3 shows the original example area (Fig. 12.1) divided into such a set of triangles.

Calculation of the area of each of the triangles in the example proceeds as follows. Consider any arbitrary triangle, as in Fig. 12.4a, where the position of each corner is defined by a pair of X- and Y-coordinates. The lengths of its sides d_1, d_2 and d_3 can be determined from the coordinates of the ends of each side using (12.4).

For any triangle of which the lengths of three sides are known, the standard trigonometric function known as the 'cosine' or 'cos' rule can be used to relate the lengths of those sides to the sizes of the angles of the triangles. So, in the example, if the angle opposite the side of length d_1 is of size θ_1, then the cos rule states that

$$d_1{}^2 = d_2{}^2 + d_3{}^2 - 2d_2d_3 \cos(\theta_1).\qquad(12.5)$$

This function can be rearranged and solved for θ_1 as

$$\theta_1 = \cos^{-1}\left[(d_1{}^2 - d_2{}^2 - d_3{}^2)/(-2d_2d_3)\right].\qquad(12.6)$$

The expression $\cos^{-1}$ in (12.6) represents the angle whose cosine is given by the

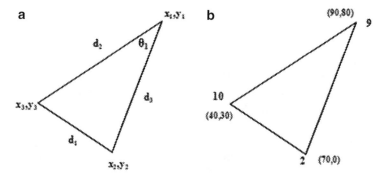

Fig. 12.4 (a) An arbitrary triangle with known coordinates of its corners and (b) the triangle defined by Points 2, 10 and 9 of Fig. 12.3. In (b), the X- and Y-coordinates (m) of each corner of the triangle are shown in *parentheses*

expression in square parentheses [ ] following it; it is called the arccosine of an angle in trigonometry. All good scientific calculators and all computer systems have functions available to determine the arccosines of angles (and their arcsines or arctangents). The sizes of the other two angles in the triangle could be determined using (12.6), with appropriate rearrangement of the positions of d_1, d_2 and d_3 in the function.

If the lengths of two sides of a triangle, say, d_2 and d_3 in the example, are known, together with the size of the angle included between them (θ_1 in the example), then the area of the triangle, A, can be determined by

$$A = d_2 d_3 \sin (\theta_1)/2. \tag{12.7}$$

Figure 12.4b shows the triangular area delimited by Points 2, 9 and 10 in Fig. 12.3, together with the X- and Y-coordinates of each corner of the triangle. Figure 12.4b has been arranged to have exactly the same form as the triangle shown in Fig. 12.4a. Applying (12.4–12.7) to the dimensions of that triangle gives $d_1 = 42.4$ m, $d_2 = 70.7$ m, $d_3 = 82.5$ m, $\theta_1 = 31.0°$ and $A = 1,500$ m^2.

If similar calculations are done for all the triangles in Fig. 12.3 and their areas summed, the total area can be calculated as 14,300 m^2, that is, 1.43 ha.

12.6 Global Positioning System

The global positioning system (usually abbreviated as GPS) allows users to determine where they are on earth (their latitude, longitude and altitude). Use of the system requires a portable electronic receiving device that can be hand-held or is easily mountable in a vehicle, boat or aircraft. The system is obviously useful as a navigational aid on land, sea or in the air but can be used also as a surveying tool.

The system is owned and controlled by the US government but is available for use by anyone who buys a suitable receiver.

The system works through 24 satellites that are positioned in earth orbit. Somewhere between five and eight of them are usually above the horizon at any time as 'viewed' from any point on earth. Ground stations around the world, controlled by the system owners, track the satellites moment by moment so their positions are always known. The satellites constantly transmit radio signals about their position. These are received by the user's portable GPS receiver. As long as signals from at least three satellites are being received, the GPS receiver can calculate its position (and, hence, that of the user) on earth as latitude and longitude. If a fourth satellite is also in 'view' of the receiver, altitude above sea level can be determined also. When more than four satellites are in 'view', GPS receivers combine the information from all of them to provide more precise fixes of position.

A GPS position estimate is accurate to about 20 m horizontally and 28 m vertically. This is more than adequate for general navigation purposes, but deviations as large as 20 m horizontally would obviously be inadequate for the example survey described in this chapter. However, GPS precision can be improved greatly by using what are known as differential global positioning system techniques (often abbreviated as DGPS techniques). These require much more sophisticated GPS receivers placed at precisely known reference locations on earth. These reference receivers can rationalise the satellite information to make it consistent with the precisely known position of the receiver.

Communities and organisations around the world own reference receivers and transmit radio information locally about corrections necessary to the satellite information or make the information available on the internet. Some portable GPS receivers can receive this radio information directly from a reference station and use it to correct the satellite information it is receiving. This can improve its precision of position estimates to as good as ± 0.1 m. With less sophisticated receivers, the corrections must be made after returning from the field using the published information.

Use of the GPS has become ubiquitous in forestry. At the very least, it allows easy navigation through the forest that would aid sample point location in an inventory. It is also being used extensively for forest mapping purposes; in conjunction with remote sensing of forests (Chap. 13), accurate maps displaying many forest attributes can be drawn today without the need for ground surveys to be carried out. The GPS is also an important safety tool to help people avoid becoming lost and to aid in crises such as firefighting or searches. Forest canopies can interfere with the radio signals on which the system relies, although usually this simply increases the time it takes the GPS receiver to receive sufficient information to make its estimate of position.

Chapter 13
Remote Sensing

Abstract In the past, most forest measurement has involved direct measurement of trees by people working on the ground. Today there is an increasing trend to exploit the sophisticated equipment available to measure trees and forests remotely from the ground, the air or space. This chapter summarises the principles and capabilities of the main types of remote sensing equipment used presently for forest measurement. For ground measurement, it considers the use of lasers for tree stem and canopy measurement, photography and other equipment for canopy measurement and ground-penetrating radar for root measurement. Air photos and, in particular, airborne laser imagery are being used widely to map and measure individual trees and stands over large forest areas. Resolution of satellite equipment is reaching a point where individual tree measurement from space may soon become a reality.

The emphasis in this book has been on direct measurement of trees by people working on the ground, often using relatively unsophisticated equipment. However, there has been an increasing trend to exploit the more sophisticated equipment now available, such as satellites, to measure trees and forests remotely, that is, without the need for people to handle the measuring instruments or, perhaps, to even visit the forest site where measurements are to be made.

Use of this type of equipment has been limited in the past, perhaps because it could not measure fully the sizes of individual trees. So rapid has been technological development that this limitation is now starting to be overcome. Over the next few decades, it is conceivable that much of the relatively labour-intensive measurement of individual trees on the ground will be superseded by the use of sophisticated electronic equipment that measures trees remotely.

This final chapter summarises the principles and capabilities of the main types of remote sensing instruments available presently. These instruments offer opportunities to view forests, map them, assess their health (e.g. Nakajima et al. 2011; Barton 2012) as well as obtain direct measurements from them (Sanchez-Azofeifa et al. 2009) and even aid in management practice (e.g. Peuhkurinen et al. 2007; Briggs et al. 2008; Wood et al. 2008; Alam et al. 2012; Watt et al. 2013b; Pierzchala et al. 2014). Different instruments are capable of operating at different scales, from the single tree right through to large forest areas. Some operate from the ground,

© Springer International Publishing Switzerland 2015
P.W. West, *Tree and Forest Measurement*, 3rd edition,
DOI 10.1007/978-3-319-14708-6_13

some are carried aloft in aircraft, whilst satellites operate from space. In this chapter, they will be discussed in that order.

13.1   Ground Measurement

Measurement of tree stems and the amount of wood they contain has been a principal focus of this book. One of the principal limitations of remote sensing instruments borne in aircraft or on satellites (Sects. 13.2 and 13.3) is that the forest canopy obscures the stems (and branches) of the trees when viewed from above; this has been an important limitation to their use more generally in forest measurement. However, instruments that are positioned on the ground are now becoming available to allow measurement of the fine detail of individual trees remotely.

13.1.1   Tree Stems and Crowns Using Lasers

Available now are ground-based, remote sensing instruments that use the reflection of laser light to construct a three-dimensional image of the trees in a stand. They operate to a distance of some tens of metres around a central point. From the information these instruments provide, it is possible to determine, in detail, the size of each tree stem as well as other above-ground parts of the tree.

The principles of laser light and its use to determine the position of a distant object was discussed in Sect. 2.2. In the context of remote sensing, laser measurement is often termed 'lidar', an acronym for LIght Detection And Ranging; this is analogous to the much older and well-known 'radar', an acronym for RAdio Detection And Ranging that uses radio, rather than light, waves.

The laser instruments being used for ground measurement of trees emit pulses of laser light that shine a spot smaller than 10–15 mm in diameter on an object (the size of the spot increases the further away the object is). This means the three-dimensional position of objects as small as leaves can be measured.

Hopkinson et al. (2004) tested one of these instruments by measuring the trees in 0.12 ha square plots in each of a mature red pine (*Pinus resinosa*) forest and a complex, uneven-aged, deciduous hardwood forest dominated by sugar maple (*Acer saccharum*) in Ontario, Canada. Their instrument could emit 2,000 laser pulses per second; the light spots were spaced apart by as little as 10 mm as the instrument scanned the three-dimensional space around it. They took six views of each plot, from points positioned outside the plots, to ensure each tree in a plot could be 'seen' clearly by the instrument. It required about 6 hours to take these views.

Complex computer programs are required to deal with the enormous amounts of data obtained from such instruments; in Hopkinson et al.'s case, they would have accumulated data for the positions of over 30 million separate points within their two plots. When these data are analysed by the computer, they can be presented as what appears as a high contrast black and white photograph showing quite clearly the stem of each tree and its branches and leaves. Hopkinson et al. used the data to determine the diameter at breast height of each tree in their stands and its total height. They found the instrument gave unbiased estimates of tree stem diameters with an accuracy quite adequate for normal forest measurement purposes. However, tree heights were underestimated by about 1.5 m on average. Because the view of the tip of a tree is largely obscured by intervening foliage, relatively few laser pulses reach the tip to be reflected back by twigs and foliage there; this 'shadowing' of the tip led to the bias in estimates of tree heights.

Henning and Radtke (2006) tested a similar instrument in a 20-year-old plantation of loblolly pine (*Pinus taeda*) in Virginia, USA. They were able to measure successfully how diameter changed along individual tree stems to a height well within the tree crown. They were able to determine also the position of branches in the lower part of the crown. However, a small degree of bias was evident in their results. Others have also measured individual tree stem profiles with such instruments (Liang et al. 2014).

Danson et al. (2007) used both a laser instrument and a camera to take the same view, from below, of the canopy of a pine forest in Switzerland. Both images are shown in Fig. 13.1 where it is obvious how closely the laser image resembles a photographic image. In their case, Danson et al. were aiming to use the laser data to determine leaf area index of the forest (that can be done also from the photographic image—Sect. 13.1.2). Other researchers have used similar laser instruments to obtain images of tree crowns (Lovell et al. 2003; Jupp et al. 2008; Seidel et al. 2012).

Tanaka et al. (2003, 2004) tested a laser instrument that operated on a slightly different principle. Instead of emitting pulses of laser light, their instrument projected a continuous laser beam that was moved progressively around the surrounding forest. Reflections from this beam were detected by a digital camera, positioned some distance from the laser instrument. Tanaka et al. were interested particularly in measuring the leaf area index of the forest canopy (see also Sect. 13.1.2). To do so required leaves to be distinguished clearly from branches and the stem in the image obtained of the canopy. They found they could do this by using laser light of two different wavelengths; leaves could be identified more clearly using infrared light, whilst branches could be better identified when lit with visible red light. Figure 13.2 shows the clear distinction between leaves, branches and stems in an image obtained by Tanaka et al.

These ground-based, laser measurement instruments are clearly showing potential for detailed measurement of tree characteristics (van Leeuwen et al. 2011). One obvious limitation is that they can only determine stem measurements over bark; if under bark measurements are required, assumptions will need to be made about bark thickness (cf. Sect. 5.4). Major electronics firms now manufacture suitable instruments that can be carried into the field and are battery operated. These instruments have many uses beyond tree measurement wherever three-dimensional views of solid objects, such as buildings, or hollow objects, such as caves, are required.

Fig. 13.1 A vertical view from below, through the canopy of a forest in Switzerland, dominated by mountain pine (*Pinus mugo*), but with some stone pine (*Pinus cembra*). Average height of the trees in the forest was about 12 m. The views are (**a**) as seen using a laser instrument and (**b**) as seen in a photograph taken with a camera with a wide-angle lens (from Fig. 2 of Danson et al. 2007, reproduced with kind permission of the Institute of Electrical and Electronics Engineers, © 2007 IEEE)

13.1.2   Leaf Area Index Using Sunlight

Leaf area index is an important stand parameter, useful to determine how much sunlight a stand absorbs and, hence, what the photosynthetic production of a stand might be. Considerable effort has been made to develop methods to measure leaf area index from the ground without having to fell trees. These have been reviewed by Fournier et al. (2003). Discussion here will be restricted to the use of instruments

Fig. 13.2 A view through the canopy of a mixed-species, hardwood forest, dominated by Japanese oak (*Quercus serrata*) trees, with an average height of 20 m, on the campus of Nagoya University, Japan. The image was derived from laser scanning of the canopy, using light of both infrared and visible red wavelengths. The *lighter elements* in the image are leaves, whilst the *greyer elements* are tree stems and branches. The information used to produce this image can be used also to determine the sizes of the various objects in the view (reprinted from Fig. 11 of Tanaka et al. 2004 with permission from Elsevier)

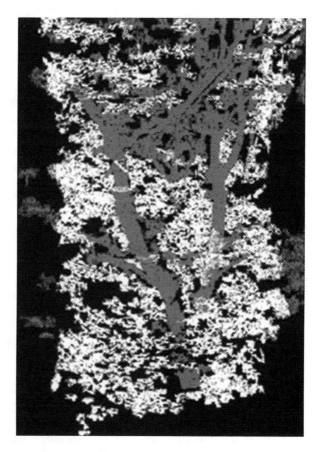

that determine leaf area index by measuring the amount of sunlight that passes through the forest canopy to the ground below.

In essence, these instruments consider the straight beams of sunlight, coming from any point in the sky above, as 'pointers' that are being projected through the canopy. The path of any beam may be interrupted, by hitting a leaf so that it does not reach the ground below, or it may pass right through the canopy and reach a measuring instrument on the ground.

By measuring how many beams of light pass through the canopy, these instruments determine the canopy gap fraction (in essence, the proportion of the area of the sky above the canopy that is not hidden by any part of the canopy). If the canopy gap fraction is known together with the angle from the horizontal at which the leaves in the forest hang, a well-known law of physics (called the Beer-Lambert law) can then be used to calculate the leaf area index of the canopy. Unfortunately, this law requires that the leaves be randomly positioned within the canopy. This is generally not the case; leaves often occur in clumps on individual shoots and shoots are often clumped in different positions within the crown. As well, leaves are not opaque and a small amount of the light that hits them passes through them. In addition, some light beams are interrupted by tree stems or branches, rather than by leaves, and some light beams are scattered by reflection from several leaves. Various methods are used to allow for these complications in measuring leaf area index with these instruments (Fournier et al. 2003; Jonckheere et al. 2005a).

Perhaps the most reliable way to allow for these complications is to calibrate the instrument specifically for the type of forest in which it is to be used. This involves determining the actual leaf area index of the canopy of a particular forest type by felling trees and directly measuring the area of their leaves. These results are compared with estimates made using the instrument, and a leaf area index estimation function is then determined by relating the actual leaf area index to that estimated by the instrument.

Dovey and du Toit (2006) did this in young plantation stands of flooded gum (*Eucalyptus grandis*) in South Africa. They used the LAI-2000 plant canopy analyser instrument. This has a wide-angle lens (these are also called fish-eye or hemispherical lenses, because they have an angle of view as wide as 180°) that directs sunlight passing through the canopy to one of five light detectors, depending on the angle above the horizon from which the light beam was directed. It detects light only in the ultraviolet to blue wavelengths. It is used on overcast days or at dawn or dusk, so sunlight is received from all angles of the sky, rather than being dominated by light beams directly from the sun. Using readings obtained from the instrument, Dovey and du Toit were able to develop a reliable leaf area index estimation function for their plantations. However, they found that the canopy characteristics of flooded gum plantations changed sufficiently, even between 2- and 3-year-old plantations, that separate functions were required for both ages. This emphasises how important it is to undertake the calibration process for any particular forest type in which the instrument is to be used. Using the same instrument, Battaglia et al. (1998) undertook a similar calibration task in 6–7-year-old plantations of shining gum (*Eucalyptus nitens*) scattered in various locations across Tasmania, Australia. Despite the leaf area indices of the different plantations varying widely (over the range 0.5–7.5 m^2/m^2), they found that a single leaf area index estimation function could be used satisfactorily for all these plantations.

Another instrument of this type is known as the DEMON. It is carried along a transect below the forest canopy whilst being pointed directly at the sun. It measures the change in light intensity received on the ground from point to point along the transect. Lang et al. (1991) tested this instrument in a plantation forest of radiata pine (*Pinus radiata*) in Canberra, Australia, and found a very close correlation between the leaf area index of the forest and the amount of light transmitted through the canopy and measured by the instrument. Lang and McMurtrie (1992) used the same instrument to measure the area of the leaves of individual trees, by moving the instrument around the ground area below a tree on which the shadow of its crown was cast. This could be a useful method to measure individual tree crowns in open forests where the trees are spaced widely apart; an instrument that measures light over a wide angle of view, such as the LAI-2000 plant canopy analyser, would be less suited for this purpose.

Both these methods of leaf area index estimation that is using diffuse light from all angles of the sky and direct beams of light from the sun were compared by Chason et al. (1991) and by Fassnacht et al. (1994). They were able to develop suitable calibrations for both pine and hardwood forests in the USA to allow satisfactory estimation of leaf area index with either method.

Another approach to measuring leaf area index from the ground is to take a photograph of the canopy, usually with a wide-angle lens, looking vertically upward from the ground below (Fig. 13.1b). Indeed, wide-angle photography has been used for many decades to obtain information on canopy characteristics. Because they provide a visual record, photographs can be useful also in assessing other things, such as damage to the canopy by insects, storms or disease.

In the past, analysis of photographs was often done by eye and could be very time consuming. For example, Koike (1985) determined the canopy gap fraction in his photographs by superimposing a fine, dotted grid over them and counting the number of points where sky or leaves appeared on each photograph. More recently, digitising (i.e. conversion to a digital form) photographs taken on film (Chan et al. 1986; Rhoads et al. 2004) and the advent of digital cameras, which provide photographs immediately in digital form, allow more rapid and thorough computer analysis of photographs. Careful selection of the sky conditions, camera settings and even the wavelengths of light detected by the camera is essential to obtain suitable contrast in the photograph between the sky and canopy elements; only then can the computer determine readily variables such as canopy gap fraction (Fournier et al. 2003; Jonckheere et al. 2005a, b, c; Zhang et al. 2005; Cescatti 2007; Chapman 2007).

Considerable research has been undertaken recently on the use of digital photography to measure leaf area index and other canopy variables. For example, Macfarlane et al. (2007b) found that digital photography, with or without a wide-angle lens, or the LAI-2000 plant canopy analyser all gave very satisfactory estimates of leaf area index in plantation stands of jarrah (*Eucalyptus marginata*) in Western Australia. Keane et al. (2005) compared wide-angle digital photography, the LAI-2000 plant canopy analyser and several other instruments that have not been mentioned otherwise here to estimate the canopy bulk density (biomass of foliage and twigs that will burn readily in a forest fire, per unit volume of the canopy) of coniferous forests in western USA. Both the LAI-2000 and wide-angle photography were found satisfactory for this.

There are many other examples of the use of wide-angle photography and other methods to measure leaf area index and other canopy variables remotely from the ground (Coops et al. 2004a; Jonckheere et al. 2004; Weiss et al. 2004; Arias et al. 2007; Macfarlane et al. 2007a, 2014; Montes et al. 2007, 2008; Schleppi et al. 2007; Wilson and Meyers 2007; Davi et al. 2008; Demarez et al. 2008; Dutilleul et al. 2008; Parveaud et al. 2008; Ryu et al. 2010; Brusa and Bunker 2014).

13.1.3 *Roots*

Perhaps the measurement of roots is the last frontier of remote sensing of forest characteristics from the ground. The excavation of roots to measure directly their biomass, length or distribution down the soil profile is an extremely labour-intensive and difficult task (Sect. 7.2.3). The development of techniques and instruments to measure roots without the need for excavation is obviously highly desirable.

One technique that shows some promise is the use of ground-penetrating radar. This technique is used commonly by engineers to locate pipes or cables that have been laid underground or by archaeologists to locate historical artefacts that have been buried for centuries or even millenia. It involves transmitting radio signals down through the soil and recording the times for reflections to be received back from objects within the soil. The higher the energy of the radio waves used (i.e. the higher their frequency), the deeper within the soil can they penetrate, perhaps to a maximum of about 10 m.

There are a number of difficulties with using radar in soil, difficulties that do not exist with the use of radar through the air (Guo et al. 2013a). The speed of travel of radio waves in air is the same as the speed of light, but soil slows that speed considerably, perhaps by more than one half; the speed is affected particularly by the temperature and amount of water in the soil (Butnor et al. 2003). This means that a ground-penetrating radar instrument must be calibrated before it is used on any day to determine the speed of travel of radio waves in the soil being considered. As well, soils contain many irregularities, such as rocks scattered throughout them, or they may have various layers with rather different properties. These irregularities can lead to unwanted 'background' reflections of radio waves. Using complex computer programs, these have to be removed from the data collected by the instrument to leave only reflections from the objects it is desired to identify.

An example of the use of ground-penetrating radar to measure root systems in forests comes from Butnor et al. (2003). They attempted to estimate the biomass of the root system to a depth of 30 cm below ground in a 34-year-old experimental plantation of loblolly pine (*Pinus taeda*) in Georgia, USA. Butnor et al. tested their system in different parts of the experiment where the growth of the trees, hence their root biomasses, had been affected substantially by the experimental treatments. Figure 13.3 shows images they derived from their data after computer manipulation to remove background reflections. It shows the distribution of roots down through the soil profile along transects in two plots of the experiment. In this case, fertilisation had led to a substantial increase in root biomass.

Butnor et al. identified a number of limitations of their technique. They were able to identify roots only with diameters greater than about 5 mm. This would exclude fine roots. Whilst fine roots usually make up only a small proportion of the total biomass of the root system (Sect. 7.4.4), they are a very important part of the physiological processes of a tree. Clearly, ground-penetrating radar does not yet have sufficient resolution to measure these very small roots. Butnor et al. found also

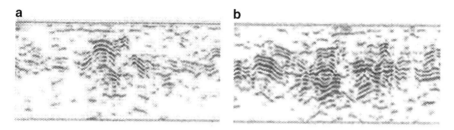

Fig. 13.3 Images, derived from a ground-penetrating radar, of the distribution of roots to a depth of 30 cm in the soil below a 34-year-old experimental plantation of loblolly pine (*Pinus taeda*) in Georgia, USA. Results are shown for (**a**) an unfertilised and (**b**) a fertilised plot in the experiment. Each image represents readings taken along a 2.4 m long transect that straddled a row of trees. The *grey markings* are the reflections from the roots of the radar signal; their relative darkness can be used to estimate the biomass of the roots at any point. The root biomass in the fertilised plot was obviously much greater than in the unfertilised. There tends also to be a greater biomass of roots near the base of a tree that was positioned in the tree row at the *centre* of each image (derived from row C of Fig. 2 of Butnor et al. 2003 and reproduced with the kind permission of the Soil Science Society of America)

that they could not measure tap roots, that is, woody roots, that grow more or less vertically immediately below the tree, often to considerable depth. They suggested that these roots might be 'seen' with the ground-penetrating radar if the instrument was oriented obliquely, rather than horizontally, to the ground. They found also that the ground surface over which the instrument was used had to be quite smooth and free of debris; this would pose a problem to the use of the instrument in native forests where understorey plants and various sorts of ground debris are common.

Other tests have been done over the last decade or so of the use of ground-penetrating radar to measure and map root systems of trees. Guo et al. (2013a) reviewed this work and concluded that

> [ground penetrating radar] has been proved to be a valuable technique for detecting coarse roots in low moisture and electrically-resistive soils. However, the detection and quantification of coarse roots ... is still in its infancy and not all roots or soil conditions are suited for this technology .... Presently, only the coarse roots that grow laterally and distribute in shallow subsoils (generally <1 m depth) and with enough moisture contrast with the surrounding soils can be detected and measured .... The biomass of root clusters can be estimated ..., but the identification of each closely-spaced, individual root cannot be accomplished .... [S]andy soils with low concentrations of organic matters and soluble salts are the most suitable condition[s] ... Overall, successful [ground penetrating radar]-based coarse root investigation is site specific, and only under suitable experiment [al] conditions can reliable measurements be accomplished.

This makes it clear that ground-penetrating radar has some potential, at least to measure woody root systems in forests. However, at present it cannot measure fine roots. Research in the use of ground-penetrating radar for root measurement is continuing (Guo et al. 2013b; Borden et al. 2014; Wu et al. 2014; Zhu et al. 2014).

A second technique that may have use in forest science and that may be able to measure fine and coarse root development within soil is X-ray computed tomography (known in medical science particularly as CAT scanning). This technique measures the attenuation of X-rays as they pass through materials and can identify soft tissues such as fine roots. However, it can be used only to detect roots growing in a small container that can be viewed by the CAT scanner from many directions. It cannot be used to measure roots below ground from the soil surface. Mooney et al. (2012) have summarised developments with this technology for root measurement in the plant sciences. Neutron radiography is another possible technique that may have application for root detection in soil; it is also only at an early stage of development for this purpose (Moradi et al. 2009).

Another technique that may have use in forest science involves measurement of the magnetisation properties of soil. This has been used to identify certain soil types in plant ecological studies (Wang et al. 2008a).

13.2   Airborne Measurement

Because of the costs involved in employing measurement crews, measurement of forests from the ground can be expected only to provide information about a limited number of individual stands within the forest. If broad-scale measurements or observations are to be taken over hundreds or thousands of hectares of forests, it is practical to do so only using instruments carried aloft in aircraft or satellites. These can provide information useful for various purposes, including identifying and mapping different forest types, assessing their site productive capacity (Sect. 8.7), locating areas of forest damaged by pests or diseases (e.g. Hanssen and Solberg 2007; Stone et al. 2013), stratifying the forest or providing auxiliary variable variables for inventory (Sects. 10.2–10.5).

As mentioned at the start of Sect. 13.1, a principal limitation to forest measurements taken from aircraft or satellites is that the forest canopy conceals the tree stems. This prevents direct measurement of stem sizes and, hence, the wood volumes they contain. Perhaps an exception to this is the possibility of measuring deciduous forest during winter when they have lost their leaves and their stems can be seen directly. An example of this was given for aerial photographs of pedunculate oak (*Quercus robur*) forest in Denmark (Tarp-Johansen 2002a, b). However, the limited number of forest types to which this applies means that little research effort has been spent on making full use of this type of imagery.

This section is concerned with the measurement of forests using instruments carried in aircraft. Different aircraft vary considerably in the speeds and altitudes at which they fly, determining ultimately the ground area they cover in any given time and the degree of resolution of the images of the forest that they produce. Fixed-wing aircraft are used most commonly, although helicopters may be used sometimes also. The potential of small remotely controlled aircraft (drones) to obtain high-resolution images from low altitude is now being recognised also (Paneque-Gálvez et al. 2014; Pierzchala et al. 2014).

13.2.1   Aerial Photography

Photographs taken from the air have been used extensively for forest management purposes for many years. Not only can aerial photographs provide measurements of some tree and stand characteristics, but they can be used also for general mapping and for vegetation studies, perhaps identifying where different vegetation types occur across the landscape or where insect attack or disease has damaged the forest.

Of recent times, it has been felt that the revolution in various forms of digital measurement (such as laser scanning—Sect. 13.2.2) may lead to the demise of aerial photography as an important tool in forest measurement (Hall 2003). Digitally measured data are particularly amenable to computer analysis and it was felt that this might avoid the need for assessments by the people who have viewed and interpreted aerial photographs in the past. However, the advent of digital cameras and the ability to transform photographs taken on film to digital form, using scanners, have largely removed this objection. As well, technological developments continue in both cameras and film that provide ever-increasing quality and variety of photographic images (Hall 2003). Also, there is information captured on photographs that the human eye is able to assess better than is possible presently with computers; hence, there remains a role for air photo interpreters. For all these reasons, aerial photographs are continuing to have a major role in mapping forests and in forest measurement.

All of black and white, infrared and colour films are used for different purposes in aerial photography; digital cameras also can take photographs using different parts of the light spectrum. Within film types, there are many subtle variations that react in different ways to light of varying wavelengths; the user may select film and camera types to suit particular needs (Hall 2003). For example, infrared photographs have been found to allow better distinction between crowns of hardwood and conifer trees in mixed forests in the northern hemisphere or to identify forests that are suffering ill health from disease (Myers et al. 1984; Avery and Burkhart 2002). The different colours of crowns of different species can aid their identification in colour photographs, even in very complex forests, such as tropical rainforest (Myers and Benson 1981).

The scale at which photographs are taken determines the resolution with which things can be seen on the ground. Both the focal length of the camera used and the altitude at which the aircraft flies determine the scale. Small-scale photographs (1:30,000–1:100,000, that is, a distance of 1 cm on the photograph represents 300–1,000 m on the ground) are suitable to recognise broadly different types of vegetation. Major types of forest cover and the species present in single-species stands can be identified from medium-scale photographs (1:10,000–1:30,000), whilst individual trees can be seen on large-scale photographs (1:2,500–1:10,000) (Avery and Burkhart 2002). Viewing the vegetation and the topography of the land on which it occurs can often be aided considerably by using pairs of photographs that overlap partially in their view of the ground; the photo interpreter can then obtain a three-dimensional image by viewing a photo pair with a stereoscope.

To map forests adequately from aerial photographs requires considerable skill and experience by photo interpreters. They need to have knowledge of the tree species and forest types that occur in the region of interest, as well as the ecological relationships within the forests and the environmental circumstances of the region. These give the interpreter clues as to what can be expected to appear on photographs. Such clues, together with characteristics such as the size, shape, colour and texture of tree crowns, will all help the interpreter to map different forest types or forest areas affected by disease or damaged by insect attack and so on; at present, the human eye is far superior at doing this than is the computer. Such maps often form the basis for the conduct of forest inventory over large, complex forest areas (Sect. 11.2).

Two examples will illustrate some uses of aerial photography. The first is from Massada et al. (2006) and concerns the estimation of the above-ground biomasses of individual trees in 40-year-old plantations of Aleppo pine (*Pinus halepensis*) in Israel. Massada et al. had available medium-scale (1:13,000) aerial photographs of the plantations. Because trees had been removed regularly from the plantations by thinning, the trees were well spaced so that their crowns could be clearly identified in the photographs. Photo interpreters were able to determine the height of individual trees by using stereo pairs of photographs and special equipment that allowed measurement of the three-dimensional coordinates of the tip of each tree and the ground below. The diameter of the crown of each tree was measured also. Massada et al. did this for each tree in a set of plots, trees that were then measured also from the ground. They found negligible differences between the ground measurements and the measurements obtained from the aerial photographs.

When ground measurements are taken, it is commonly the tree diameter at breast height and tree height that are measured. Provided a biomass estimation function is then available for the species concerned, a function such as (7.1), individual tree biomasses can then be determined. However, because the forest canopy hides the tree stems, it is not possible to measure stem diameters from aerial photographs. Massada et al. dealt with this problem by developing a new biomass function for their species from ground measurements of biomasses, crown diameters and heights of a set of sample trees. Their function was

$$B_A = 0.259 C^{1.48} H^{1.67} \tag{13.1}$$

where B_A was above-ground oven-dry biomass (kg), C was crown diameter (m) and H was tree height (m). It can be argued that larger trees will tend to have wider crowns and so tree crown diameter should be correlated quite highly with stem diameter. Thus, (13.1) can be considered as an allometric biomass estimation function, where crown diameter replaces stem diameter at breast height and that also includes tree height (cf. Sect. 7.3). This function could now be used by Massada et al. to estimate biomasses of individual trees from the tree measurements they were able to take from their aerial photographs. Methods such as this are used commonly with various methods of remote sensing in forests. Where the remote sensing method is unable to provide some measurement that would be taken normally from the ground, a method is developed to provide the result desired ultimately based on those measurements that the remote sensing is able to provide.

The second example is from Harcombe et al. (2004) and concerns a problem in the native forests of mixed Sitka spruce (*Picea sitchensis*) and western hemlock (*Tsuga heterophylla*) in the mountainous coastal ranges of Oregon in north-western USA. This region is subject to intense wind storms that blow in from the sea during winter and may be sufficiently strong to blow down patches of forest.

Harcombe et al. had available six sets of medium-scale, aerial photographs that covered some 500 ha of the region and that had been taken in each of various years between 1953 and 1993. Using stereo pairs of the photographs, photo interpreters were able to see and map patches of forest, as small as 0.5 ha in area, where tree stems were lying on the ground after being blown down by the wind. The difficulty of identifying such patches on a photograph is probably a good example where the human eye would be superior to the present capabilities of a computer.

In 1953, Harcombe et al. could identify only two small areas of blown-down forest with a total area somewhat less than 5 ha. However, over the next 40 years, additional blown-down areas appeared and the existing areas expanded progressively and coalesced. By 1993 all these areas formed a single, large patch of nearly 50 ha of blown-down forest. Harcombe et al. were able to correlate the location of this patch with the condition of the forest and the topography and wind intensities of the region. They concluded that the risk of forest blowdown was greatest when the forest was at least 100 years old, by which time the trees had grown to an average height of about 50 m, and where the forest was growing on particularly exposed southerly facing slopes of secondary ridges to the west of the main north-south ridge of the mountains of the region. This information would be useful in making decisions about appropriate management and conservation practices for these forests.

The difficulty of the terrain in this region and the long time period involved in the development of the large blown-down patch would have made it virtually impossible to have accumulated ground measurement information to give these results. It was only the availability of the long-term set of aerial photographs that allowed Harcombe et al. to do so. There are some interesting examples of the use of long-term sets of aerial photographs to study vegetation change (Fensham and Fairfax 2002; Fensham et al. 2002; Witt et al. 2009).

13.2.2 Laser Scanning

Laser scanning uses laser light transmitted from an aircraft or a satellite, some of which is reflected back when it strikes a solid object on the ground below (Holopainen et al. 2014). This is another application of lidar (Sect. 13.1.1).

Usually for this form of remote sensing, a laser is used that emits 50,000–100,000 pulses per second that reach the ground as spots. The size of the spots and the distance along the ground of successive pulses varies with the speed over

Fig. 13.4 A cross-sectional view of the canopy of a eucalypt forest in southeast Queensland, Australia, derived using laser scanning information from an aircraft flying at about 250 km/h at an altitude of about 1,200 m. The total distance across the image was about 500 m; the vertical and horizontal scales are the same. The *solid line* at the base of the image is the ground surface that rises gently from right to left and then falls away quite sharply near the *left-hand end* of the image. The *dotted points* represent the top of the vegetation canopy along the cross section, measured in this case at about 1 m intervals; a contiguous set of 6–8 points in the upper reaches of the canopy might represent the crown of a single tree. The average height of the trees along this cross section was measured on the ground as 24 m (from MBAC Consulting 2003, © Commonwealth of Australia, reproduced by permission)

the ground and altitude of the aircraft or satellite and the quality of the laser equipment.

When the object on the ground being viewed is not completely solid, such as is the case for the canopy of a forest, reflections are obtained from the light spots as they strike various solid objects (leaves, branches, understorey vegetation or the ground below) during their passage down through the canopy. The tiny time differences between these multiple reflections can be measured so the heights of various objects within the canopy can be determined; differences in distances as small as 0.1–0.3 m can be measured using this method. The first reflections received will be from the top of the canopy and the last will be from the ground below the canopy, so that the height of the trees in the forest can be determined. An example of the type of the image of the forest canopy that can be obtained is illustrated in Fig. 13.4. St-Onge et al. (2003) and Lovell et al. (2005) have reviewed the issues surrounding tree height estimation using airborne lidar.

With laser data collected at sufficiently fine resolution, say with a spot size of about 0.25 m and with spots located 0.5–1 m apart, the position, height and crown spread of individual trees in stands can be determined readily. Considerable research attention has been paid to the application of airborne laser scanning to attempt to use the data to estimate many other individual tree and stand characteristics. As with aerial photographs, this often requires development of functions that relate those characteristics to the variables that are measured directly by the laser scanner.

Airborne lidar has rapidly become widely used for forest measurement. Examples of its use in various forest types around the world are for the measurement of tree stem diameters, heights, stem volumes and crown structure, stand frequency distribution of diameters, wood volume, biomass above and below ground, tree wood properties, woody debris and overstorey and understorey structure and for broad-scale forest inventory (Persson et al. 2002; Holmgren et al. 2003; Lovell et al. 2003; Coops et al. 2004c, 2007; Maltamo et al. 2004, 2005, 2006a, b, c, 2007a, b; Næsset 2004, 2007; Popescu and Wynne 2004; Popescu et al. 2004; Riaño et al. 2004; St-Onge et al. 2004; Gobakken and Næsset 2005, 2008; Hall et al. 2005; Roberts et al. 2005; Chasmer et al. 2006; Falkowski et al. 2006; Mehtätalo 2006; Tickle et al. 2006; Bollandsås and Næsset 2007; Hanssen and Solberg 2007; Magnusson et al. 2007; Peuhkurinen et al. 2007; Popescu 2007; Bollandsås et al. 2008; Breidenbach et al. 2008; Briggs et al. 2008; Fujisaki et al. 2008; Heurich 2008; Hilker et al. 2008; Pascual et al. 2008; Pesonen et al. 2008; van Leeuwen et al. 2011; Hyyppä et al. 2012a, b; Watt et al. 2013a; Cao et al. 2014; Hayashi et al. 2014; Vincent et al. 2014; Saremi et al. 2014a, b; Schreyer et al. 2014). Attempts are now being made to adapt aerial photography to provide similar information to airborne laser scanning at rather less cost (White et al. 2013).

13.2.3 Spectrometry

A spectrometer is an instrument that records the amount of light it receives at each of a very wide range of wavelengths across the radiation spectrum. Typically, it might record the light received from as many as 300 separate, narrow wavelength bands in the visible or infrared light regions. In this context, a spectrometer is similar to a camera, except that a camera produces an image that combines the light received at many wavelengths, whereas a spectrometer records separately the light received at each wavelength.

Spectrometers can be used on the ground, from the air or can be carried in satellites. However, for forestry purposes there are some good examples of their use when carried in aircraft. Just as with aerial photographs, the properties of the instrument and the altitude at which the aircraft flies will determine the scale on the ground of the spectrometer recordings. At sufficiently large scale, they can certainly record the radiation reflected from the crowns of individual trees on the ground below.

An example of the use of spectrometry concerns assessment of the health of both native eucalypt forest and plantations of radiata pine (*Pinus radiata*) in New South Wales, Australia (Coops et al. 2003a, b, 2004a). One common symptom of the ill health of trees is a reduction in the concentration of chlorophyll in their leaves; chlorophyll is a green pigment contained in leaves and is a crucial part of their photosynthetic system. Because chlorophyll absorbs light in red wavelengths strongly, it might be expected that a tree in poor health with a low chlorophyll content in its leaves would not absorb red light as strongly; a spectrometer might then be used to identify trees that are absorbing relatively low amounts of red light.

Using an airborne spectrometer, Coops et al. (2003b) were able to identify trees in a native eucalypt forest that had low chlorophyll contents as a result of damage to their crowns by leaf-eating insects. Coops et al. (2004b) found they could use the information from other wavelengths to identify directly individual trees that had damaged crowns. In the same forest area, Goodwin et al. (2005) found they could use the spectral characteristics of individual tree crowns to discriminate non-eucalypt trees from eucalypts but were unable to separate eucalypt species one from another. Similarly, in a radiata pine plantation, Coops et al. (2003a) found they could identify individual trees of which the crowns had been damaged and discoloured by a disease known as Dothistroma needle blight, a fungal disease that causes leaves to be shed from trees.

Another form of airborne spectrometry measures the concentrations of three elements, potassium, thorium and uranium, in the top 35–40 cm of the soil on the ground below. It does so by measuring the emission of γ-rays (radiation of a rather short wavelength) emitted during radioactive decay of these elements. This can be used to infer various properties of the soil including its depth, texture and its degree of weathering. Wang et al. (2007b) used the information from such a spectrometer as a part of a system to predict the site productive capacity (Sect. 8.7) of pine plantation forests across wide areas of Queensland, Australia.

13.3   Satellites

With their worldwide coverage at all times of year, satellites offer one of the most comprehensive forms of remotely sensed information (often referred to as satellite imagery) from forests. Some satellites are passive, that is, they sense radiation reflected from the surface of the earth. Others are active, that is, they emit radio or laser radiation that is reflected from the surface below back to the satellite.

As an example, the NASA-US Geological Survey owned Landsat satellite series is used widely for forestry purposes. There have been seven satellites in this series, launched from time to time between 1972 and 2013; two, named Landsat 7 and Landsat 8, are still functional. Between all the satellites of the series, they provide over 40 years of data offering the possibility of studying changes that have occurred over that time in the vegetation at any point on earth. Landsat records images of the earth at a number of different light wavelengths at a resolution of 30 × 30 m of the ground surface; that is, the images are made up of square spots, referred to commonly as pixels, that represent the intensity of radiation received from areas of the ground of that size. It also takes photographs at a resolution of 15 × 15 m. These resolutions are inadequate to identify or measure individual trees in a forest. However, different types of vegetation and objects on the land surface will reflect light of various wavelengths differently; it is these differences that offer the opportunity to identify and measure differences in vegetation or other land characteristics from point to point across the ground surface.

There are now a large number of satellites, both privately and government owned, that can provide information that might be useful for forest measurement. They produce images at each of several light wavelengths and, as their technology improves, at finer and finer resolution. Some are attaining a resolution that allows individual tree crowns to be identified. Figure 13.5 shows the canopy of a forest as viewed by two satellites using stereo photographic imagery and also by using airborne laser scanning. Clearly laser scanning (Fig. 13.5c) offered the best resolution of individual tree canopies in this view, but the WorldView-2 satellite

Fig. 13.5 An oblique scene of the tree canopies over an area of mixed hardwood and softwood forest (the principal species being spruce, *Picea abies*, beech, *Fagus sylvatica*, and fir, *Abies alba*) scattered in patches across the landscape in Bavaria, Germany. The scene was obtained using stereo photographic imagery from (**a**) the Cartosat-1 satellite (Indian Space Research Organisation) and (**b**) the WorldView-2 satellite (DigitalGlobe Corporation) as well as by (**c**) airborne laser scanning (lidar). The distance across the width of the view is 350 m (from Fig. 4 of Straub et al. 2013, reproduced by permission of the principal author, C Straub, and Oxford University Press on behalf of the Institute of Chartered Foresters, UK)

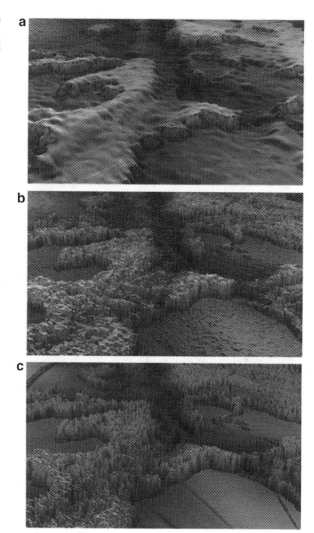

(Fig. 13.5b) is approaching that resolution. The Cartosat-1 satellite (Fig. 13.5a) has lower resolution and provided only a general surface view of the canopies. Work is presently underway to develop a high-resolution laser imaging instrument to be known as the Global Ecosystem Dynamics Investigation lidar that will be carried on the International Space Station (Zubritsky 2014); this may help to overcome the limitations apparent in the satellite imagery in Fig. 13.5. There have been satellites offering active laser scanning and these have been used for large-scale mapping of forest canopy height (Lefsky 2010) and biomass (Ballhorn et al. 2011).

To use satellite images effectively requires specialists who can, with the aid of a computer, adjust the images to take account of various technical problems associated with them. The exact area of the surface of the earth that any image covers must be identified. Allowance must be made also for matters such as the angle of view of the image, the angle of the sun above the horizon at the time the image was taken and variations in the atmosphere.

Several examples will be used to illustrate the ways in which satellite images are being used presently. The first example concerns the monitoring of young Sitka spruce (*Picea sitchensis*) plantations in Britain to determine if they have developed adequately (Donoghue et al. 2004). The average height and stand basal area of a number of plots located in 2–17-year-old Sitka spruce plantations were measured on the ground. It was found that their average height correlated well with infrared light intensity as measured for the plots from a satellite, whilst stand basal area correlated reasonably well with light intensity measured in green wavelengths. The results were similar with data collected either by the Landsat 7 or the French-owned SPOT satellites. It was concluded that the satellite information was sufficient to allow assessment of the viability or otherwise of individual plantations.

The second example is drawn from work in China (Jiang et al. 1999), where an attempt was made to assess the rate at which a wide variety of forest types, spread across the whole country, were accumulating and storing carbon through photo-synthesis. Jiang et al. had available data on the annual rate of carbon accumulation (a measure known as net primary production, often abbreviated as NPP) by over 30 different forest types, both coniferous and hardwood, measured on the ground in over 1,000 plots spread around the country. Many studies around the world have found that net primary production by vegetation is related to a measure that can be calculated from satellite data, a measure known as normalised difference vegetation index (abbreviated commonly as NDVI). It is calculated, from light intensities measured in satellite data from both red and infrared wavelengths, as

$$N = (I_r - I_i)/(I_r + I_i), \tag{13.2}$$

where N is the normalised difference vegetation index and I_r and I_i are the intensities of red and infrared, respectively, light measured in the same pixel of a satellite image. Jiang et al. obtained their images from the NOAA-AVHRR satellite, an American-owned satellite. For each forest type, they related net primary production to normalised difference vegetation index using the function

$$N_P = a[1 - \ln(1 - bN)], \tag{13.3}$$

where N_P was net primary production, a and b were parameters, the values of which varied with vegetation type, and $\ln(\cdot)$ denotes natural logarithms. Given a map showing where different forest types occurred around the country, Jiang et al. were then able to use satellite images from the whole country with (13.3) to estimate and map how net primary production of forests varied right across China.

There are many other cases where normalised difference vegetation index obtained from satellite data has been used to assess vegetation net primary production and other vegetation variables, including leaf area index, the proportion of incoming sunlight absorbed by the canopy, growth in stand stem wood volume and species richness (Gholz et al. 1991; Coops et al. 1997, 1998, 1999, 2001; Goetz et al. 1999; Waring et al. 2002; Wythers et al. 2003; Le Maire et al. 2005; Pocewicz et al. 2004; Richards and Brack 2004; Van Tuyl et al. 2005; Volcani et al. 2005; Chirici et al. 2007; Meng et al. 2007; Marsden et al. 2010).

The last example (Austin et al. 2003) concerns the use of an active radar satellite sensor, the Japanese Earth Resources Satellite. Radar is sensitive to the presence of water and it is argued that this should aid its ability to detect, hence measure, fresh plant biomass. In this example, it was found that the level of radar reflection correlated reasonably well with both the stand above-ground live tree biomass and the biomass of coarse woody debris (woody material fallen from trees or in dead, standing trees) measured on the ground in a set of plots in open eucalypt forest in southern New South Wales, Australia. It was concluded that this form of remote sensing had some potential for the estimation of forest stand biomass. Coops (2002) has discussed some of the problems involved with the use of radar in assessing forest biomass. There are numerous other examples of the estimation of forest biomass using satellite imagery (Mallinis et al. 2004; Magnusson and Fransson 2005; Hall et al. 2006; Labrecque et al. 2006; Suganuma et al. 2006; Meng et al. 2007; Michelakis et al. 2014). Others have also used satellite radar to map forest areas (da Costa Freitas et al. 2008; De Grandi et al. 2009).

Satellite imagery has been used also to map different forest types (Moisen and Edwards 1999; Frescino et al. 2001; Lu 2005; Moisen and Frescino 2002), to determine the density of forest canopies (Behn et al. 2001; Baynes 2004), to determine the age structure of forests (Drezet and Quegan 2007), to identify forest suffering decline (Wang et al. 2007a), as an aid in predicting forest growth over large regions (Nightingale et al. 2008a, b; Smith et al. 2008; Lefsky 2010; Waring et al. 2010; Ballhorn et al. 2011), to estimate rainfall interception by forest canopies (Nieschulze et al. 2009) and as an aid in forest inventory (Köhl and Kushwaha 1994; Moisen and Edwards 1999; McRoberts et al. 2006; Schreyer et al. 2014).

Appendix 1: Glossary

Words shown in **bold type** have a separate entry in the glossary.

Accuracy	The difference between a measurement or estimate of something and its true value
Allometry	The relationship between part of an organism and its whole
Ancillary information	A variable that has some degree of **correlation** with some other variable of interest
Auxiliary information	A variable that has some degree of **correlation** with some other variable of interest
Basal area	Cross-sectional area at **breast height** of a **tree** stem. See also **stand basal area**
Basic density	The **oven-dry** weight of **wood** per unit green volume
Bias	The difference between the average of a set of repeated measurements or estimates of something and its true value
Bioenergy	**Biomass** used to make energy, usually by conversion to ethanol or burning to generate electricity
Biomass	The weight of a living organism. It may include the water in the organism, when it is referred to as fresh biomass. Often, the tissue is dried and its **oven-dry** biomass is considered
Biomass expansion factor	The ratio between the **biomass** of some part of a tree and the volume of the stem of the tree. It can also be the ratio between the **stand** biomass of a tree part and the **stand** stem volume
Breast height	A height of 1.3 m (or 1.4 m in some countries and 4′6″ in the USA) above ground from the base of a **tree**. If the **tree** is growing on sloping ground, it is measured from the highest ground level at the base of the **tree**
Canopy	The foliage and branches of a **forest**

© Springer International Publishing Switzerland 2015
P.W. West, *Tree and Forest Measurement*, 3rd edition,
DOI 10.1007/978-3-319-14708-6

Confidence interval	The range, evaluated with known probability, within which the true **mean** of a **population** lies, when the mean has been estimated from a sample from the **population**. It is derived from the **confidence limit**
Confidence limit	A mathematical statistical measure to determine, with a known probability, the limits within which the true **mean** of a **population** lies, when the mean has been estimated from a sample from the **population**
Correlation	Two variables, measured on each of a set of objects, are said to be correlated when the value of one tends to change systematically with the value of the other
Covariate	A variable that has some degree of **correlation** with some other variable of interest
Crown	The foliage and branches of a **tree**
Current annual increment	The present growth rate of a **stand** at any particular age. It is often abbreviated as CAI. Also known as **periodic annual increment** (PAI)
Dendrometer	An instrument to measure **tree** stem diameter
Digital	Forms of measurement where all the information obtained is as either one of two 'digits' only. The two digits might be recorded as a zero or a one, as the presence or absence of something and so on. These digits can then constitute a code to describe something, a code that computers are particularly efficient at interpreting. The digits are quite distinct and so are unlikely to be misinterpreted, unlike analogue measurements that are continuously variable and so are more subject to misinterpretation
Dominant height	Average height of a prescribed number per unit area of the tallest or largest diameter **trees** in a **stand** (see also **top height** and **predominant height**)
Ecosystem	An assemblage of plants and animals living together at a **site**
Empirical	Based on experiments or observations, rather than on theory
Environment	The other living or inanimate things amongst which a living organism grows and reproduces
Even-aged	All the **trees** in a **stand** regenerated naturally (in **native forest**) or were planted (in **plantations**) at or about the same time. Generally there would be less than one year difference in age between any of the **trees** in the stand (cf. **uneven-aged**)
Forest	Vegetation dominated by plants with **woody** stems that reach a mature height in excess of a few metres

Forest management	Activities undertaken in a **forest** to achieve the provision of the goods and services that are desired from it
Forestry	The use and **management** of **forests** to provide goods and services to people
Functional form	The variables included, and the way they are arranged algebraically, in a mathematical function. These determine the shape(s) the function may adopt
Geographic information system	A computer system to store spatial data and draw maps. They allow mapping of physical features of the landscape and areal mapping of characteristics of interest that may be overlaid onto a single map as desired
Global positioning system	A satellite-based system, deployed and operated by the US government, that allows the user to determine where they are on earth. Anyone may purchase a hand-held, relatively inexpensive electronic device that receives satellite signals and determines their latitude and longitude and their altitude above sea level at any instant
Hardwood	**Tree** species that are flowering plants in which the seeds develop enclosed in an ovary (cf. **softwood**)
Height	The height of a **tree** is the vertical distance from ground level to the highest green point on the **tree**
Hypsometer	An instrument that is based on geometric or trigonometric principles for measuring **tree** height
Inventory	Measurement or estimation of characteristics of a large forested area
Laser	An acronym for **L**ight **A**mplification by **S**timulated **E**mission of **R**adiation. Laser light involves an intense, narrow beam of light of a single colour that can be directed very precisely
Leaf area index	The area of the leaves of a **forest canopy**, expressed per unit area of the ground they cover. Leaf area is defined as the area of the shadow that the leaves would cast if laid flat and lit vertically from above
Management	See **Forest management**
Mathematical statistics	A branch of mathematics concerned with methods to study, condense and make generalisations about information observed in natural systems
Mean	The average of the values in a set of data
Mean annual increment	The average rate of production (of **wood**, biomass, basal area, etc.) to any particular age of a **stand**. It is often abbreviated as MAI
Median	The value in a set of data that has an equal number of values above and below it

Merchantable volume	The volume of part of a **tree** stem that can be sold to convert to **wood** products by processes such as saw-milling or paper pulp manufacture
Mode	The most common value in a data set
Native forest	**Forest** that has regenerated following a disturbance (such as fire, storm or logging by man) and has been allowed to develop more or less as would happen naturally without intervention by man
Nutrient	Any one of 15 chemical elements that are essential for plants and that play a wide variety of roles in their metabolism. They are available to land plants mainly dissolved in water in the soil and are taken up by the roots. The nutrient elements required in largest amounts by plants are nitrogen, potassium, magnesium, phosphorus, calcium and sulphur
Oven-dry	Term to describe tissue after it has been dried (usually at 60–80°C) in an oven until its weight becomes constant
Parameter	A variable in an equation that takes a particular values for a particular set of measured variables, the relationship amongst which the equation is being used to define
Periodic annual increment	See **current annual increment**
Photosynthesis	The process of chemical conversion by plants of carbon dioxide, taken into their leaves from the air, to sugars that are then used to provide energy to the plant for other metabolic processes. Light absorbed by the leaves from the sun provides the energy required in this. Oxygen is released from the leaves as part of the process
Plantation	A **forest** created by man, where seeds or seedlings have been planted, usually at a regular spacing
Point sampling	A method of measuring certain **stand** characteristics from a single point within the stand
Population	A clearly defined set of things of interest
Population statistic	A measure used to summarise a characteristic of a **population**
Precision	The variation in a set of repeated measurements or estimates of something
Predominant height	Average height of a prescribed number per unit area of the tallest **trees** in a **stand** (see also **dominant height**)
Pulplog	A small log, cut from a **tree** stem, of a size appropriate for chipping to be used for making paper
Quadratic mean diameter	The diameter of the **tree** of average **basal area** in a stand

Rainforest	**Forest** that occurs in areas with high rainfall. In the tropics, rainforests contain a large number of species of tall, broad-leaved, evergreen **trees** that form a continuous **canopy**. In temperate zones, rainforests also contain many **tree** species, but are distinguished from tropical rainforests by having dominant individual species (adapted from the Merriam-Webster Collegiate Dictionary, 10th edition)
Remote sensing	Measurement methods relying on equipment that measures or detects objects at some distance from the equipment
Sample	A set of **sampling units** selected from a **population**. They will be measured and the results used to attempt to make inferences about the properties of the whole **population**
Sampling unit	A clearly defined part of, or individual in, a **population** that might be included as one member of a **sample** drawn from the **population**
Sawlog	A log cut from a **tree** stem and large enough to be sawn into one or more of the many types of sawn **wood** used for building and many other purposes
Silviculture	The tending of **trees** in **forests** to achieve some desired objectives of management
Site	A more or less homogeneous area of land, across which **site productive capacity** is more or less constant
Site index	A measure of **site productive capacity**, defined as the **top height** or **predominant height** of a **stand** at a prescribed age
Site productive capacity	The total **stand** biomass produced, up to any particular stage of development, of a **forest** growing on a particular site, when it uses fully the resources necessary for **tree** growth that are available from the site
Softwood	**Tree** species that do not have flowers and in which the seeds develop without the protection of an ovary. Often these 'naked' seeds are protected by the scales of a cone (cf. **hardwood**)
Stand	A more or less homogeneous group of **trees** in a **forest** in which an observer might stand and look about him or her
Stand basal area	Stem cross-sectional area at breast height, summed over all the **trees** in a **stand** and expressed per unit ground area
Stocking density	The number of **tree** stems per unit area in a **stand**
Stratum	A subdivision of a **population** containing **sampling units** that have characteristics in common
Taper function	A mathematical function that predicts the diameter of the stem of an individual **tree** at any distance along its stem
Timber	**Wood** cut from **tree** stems into sizes appropriate for its final use. Known as lumber in America

Top height	Average height of a prescribed number per unit area of the largest diameter **trees** in a **stand** (see also **dominant height**)
Tree	A **woody** plant with a distinct stem or stems and with a mature height of several metres
Understorey	A layer of vegetation growing beneath the main **canopy** of a forest
Uneven-aged	The **trees** in a **forest stand** are of a wide range of ages (cf. **even-aged**)
Variance	A measure of the amount of variation in a set of measurements. It is a concept derived from **mathematical statistics** and has a formal mathematical definition discussed in the text
Volume function	A mathematical function that allows estimation of the volume of the stem, or parts of the stem, of an individual **tree** from simple measurements that can be taken from the ground, such as diameter at **breast height** over bark and **tree** total **height**
Wood	A strong material forming the greater part of the stem, branches and woody roots of **trees**. It consists mainly of dead tissue
Wood density	See **Basic density**
Woodland	Open **forests** in which the **tree crowns** cover 20–50 % of the land area

Appendix 2: Conversion Factors

Abbreviations used commonly are shown in parentheses.

Metric-imperial conversion factors
1 centimetre (cm) = 0.3940 inches (in)
1 metre (m) = 3.2808 feet (ft) = 1.094 yards (yd)
1 hectare (ha) = 2.471 acres (a)
1 kilogram (kg) = 2.205 pounds (lb)
1 tonne (t) = 0.9842 tons
1 kilometre (km) = 0.6214 miles
1 litre = 0.2120 gallons (Br) = 0.2642 gallons (USA)
1 millilitre (ml) = 0.0352 fluid ounces (fl oz)

Conversions within the metric system
1 cm = 10 millimetre (mm)
1 m = 100 cm = 1,000 mm
1 km = 1,000 m
1 ha = 10,000 m^2
1 t = 1,000 kg
1 litre = 1,000 cm^3 = 1,000 ml

Conversions within the imperial system
1 ft = 12 in
1 yd = 3 ft
1 chain = 100 links = 22 yd
1 furlong = 10 chains
1 mile = 8 furlongs = 1,760 yd = 5,280 ft
1 acre = 10 chain^2 = 4,840 yd^2
1 lb = 16 ounces (oz)
1 ton = 2,240 lb
1 gallon = 4 quarts = 8 pints

© Springer International Publishing Switzerland 2015
P.W. West, *Tree and Forest Measurement*, 3rd edition,
DOI 10.1007/978-3-319-14708-6

1 super foot = 1/12 ft^3
1 cord = 128 ft^3
1 cunit = 100 ft^3

Appendix 3: The Greek Alphabet

Uppercase	Lowercase	Letter
A	α	Alpha
B	β	Beta
Γ	γ	Gamma
Δ	δ	Delta
E	ε	Epsilon
Z	ζ	Zeta
H	η	Eta
Θ	θ	Theta
I	ι	Iota
K	κ	Kappa
Λ	λ	Lambda
M	μ	Mu
N	ν	Nu
Ξ	ξ	Xi
O	o	Omicron
Π	π	Pi
P	ρ	Rho
Σ	σ	Sigma
T	τ	Tau
Y	υ	Upsilon
Φ	φ	Phi
X	χ	Chi
Ψ	ψ	Psi
Ω	ω	Omega

© Springer International Publishing Switzerland 2015
P.W. West, *Tree and Forest Measurement*, 3rd edition,
DOI 10.1007/978-3-319-14708-6

Appendix 4: Basic Trigonometry

The Right-Angled Triangle

Consider the right-angled triangle below with sides of lengths a, b and c and containing an angle of size B.

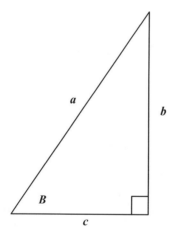

Pythagoras' theorem states that the square on the hypotenuse equals the sum of the squares on the other two sides. That is,

$$a^2 = b^2 + c^2.$$

By definition, for angle B,

$\sin(B) = $ length opposite side/length hypotenuse $= b/a$
$\cos(B) = $ length adjacent side/length hypotenuse $= c/a$
$\tan(B) = $ length opposite side/length adjacent side $= b/c$

© Springer International Publishing Switzerland 2015
P.W. West, *Tree and Forest Measurement*, 3rd edition,
DOI 10.1007/978-3-319-14708-6

Note that the abbreviations sin, cos and tan refer to sine, cosine and tangent, respectively.

Any Triangle

Consider the triangle below with sides of lengths a, b and c and containing angles of sizes A, B and C.

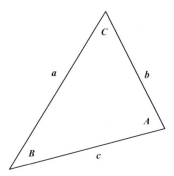

For any such triangle, $A + B + C = 180°$.
As well, provided that you know

- the lengths of all three sides, or
- the lengths of two sides and the size of one angle, or
- the length of one side and the sizes of two angles,

then the lengths of the unknown sides and the sizes of the unknown angles may be determined using the following formulae; some algebraic manipulation of these formulae may be required for you to do so.

Firstly, the 'cosine rule' determines the length of a third side if the lengths of the other two sides and their included angle are known. So, for example, if lengths b and c are known, together with the size of their included angle, A, then length a may be obtained from this rule that states

$$a^2 = b^2 + c^2 - 2bc \cos (A).$$

Secondly, the sides and angles have the properties that

$$a/\sin (A) \ = b/\sin (B) \ = c/\sin (C).$$

In addition, the area (A) of any such triangle may be determined if the lengths of any two sides and their included angle are known. So, for example, if b, c and A are known, then

$$A = bc[\sin(A)]/2.$$

Some Useful Properties of Trigonometric Functions

For any angles of sizes p and q (in degrees)

$\tan(p) = \sin(p)/\cos(p)$
$\sin(-p) = -\sin(p)$
$\cos(-p) = \cos(p)$
$\sin(90° - p) = \cos(p)$
$\cos(90° - p) = \sin(p)$
$\sin^2(p) + \cos^2(p) = 1$
$\sin(p + q) = \cos(p)\sin(q) + \sin(p)\cos(q)$
$\cos(p + q) = \cos(p)\cos(q) - \sin(p)\sin(q)$

Note that computer systems often include trigonometric functions. Generally, these require that angles be expressed in the angular measure radians, rather than degrees. The conversion factor is $90° = \pi/2$ radians ≈ 1.570796 radians.

About the Author

Professor Phil West has been a forest scientist for over 40 years. His research specialities concern the growth behaviour of forests and forest measurement. He is presently a forestry consultant and teaches forest measurement, plantation forestry and plant physiology and ecology in the forestry school of Southern Cross University in northern New South Wales, Australia.

© Springer International Publishing Switzerland 2015
P.W. West, *Tree and Forest Measurement*, 3rd edition,
DOI 10.1007/978-3-319-14708-6

References

Akindele SO, LeMay VM (2006) Development of tree volume equations for common timber species in the tropical rain forest area of Nigeria. For Ecol Manage 226:41–48

Al Afas N, Marron N, Zavalloni C, Ceulemans R (2008) Growth and production of short-rotation coppice culture of poplar—IV: Fine root characteristics of five poplar clones. Biomass Bioenergy 32:494–502

Alam MM, Strandgard MN, Brown MW, Fox JC (2012) Improving the productivity of mechanised harvesting systems using remote sensing. Aust For 75:238–245

Aleixo da Silva JA, Borders BE, Brister GH (1994) Estimating tree volume using a new form factor. Comm For Rev 73:14–17

Alvarez E, Duque A, Saldarriaga J, Cabrera K, Salas GD, del Valle I, Lema A, Moreno F, Orrego S, Rodríguez L (2012) Tree above-ground biomass allometries for carbon stocks estimation in the natural forests of Colombia. For Ecol Manage 267:297–308

Ambrose AR, Sillett SC, Dawson TE (2009) Effects of tree height on branch hydraulics, leaf structure and gas exchange in California redwoods. Plant Cell Environ 32:743–757

Annighöfer P, Mölder I, Zerbe S, Kawaletz H, Terwei A, Ammer C (2012) Biomass functions for the two alien tree species *Prunus serotina* Ehrh. and *Robinia pseudoacacia* L. in floodplain forests of Northern Italy. Eur J For Res 131:1619–1635

Antón-Fernández C, Burkhart HE, Strub M, Amateis RL (2011) Effects of initial spacing on height development of loblolly pine. For Sci 57:201–211

Arias D, Calco-Alvarado J, Dohrenbusch A (2007) Calibration of LAI-2000 to estimate leaf area index (LAI) and assessment of its relationship with stand productivity in six native and introduced tree species in Costa Rica. For Ecol Manage 247:185–193

Asner GP, Mascaro J, Muller-Landau HC, Vieilledent G, Vaudry R, Rasamoelina M, Hall JS, van Breugel M (2012) A universal airborne LiDAR approach for tropical forest carbon mapping. Oecologia 168:1147–1160

Atwell BJ, Kriedemann PE, Turnbull CGN (eds) (1999) Plants in action. Macmillan, Melbourne

Austin JM, Mackey BG, Van Niel KP (2003) Estimating forest biomass using satellite radar: an exploratory study in a temperate Australian *Eucalyptus* forest. For Ecol Manage 176:575–583

Avery TE, Burkhart HE (2002) Forest measurements, 5th edn. McGraw-Hill, New York

Baldwin VC, Peterson KD, Clark A, Ferguson RB, Strub MR, Bower DR (2000) The effects of spacing and thinning on stand and tree characteristics of 38-year-old Loblolly Pine. For Ecol Manage 137:91–102

Ballhorn U, Jubanski J, Siegert F (2011) ICESat/GLAS data as a measurement tool for peatland topography and peat swamp forest biomass in Kalimantan, Indonesia. Remote Sens 3:1957–1982

© Springer International Publishing Switzerland 2015

P.W. West, *Tree and Forest Measurement*, 3rd edition,

DOI 10.1007/978-3-319-14708-6

Banin L et al (2014) Tropical forest wood production: a cross-continental comparison. J Ecol 102:1025–1037

Baral SK, Schneider R, Pothier D, Berninger F (2013) Predicting sugar maple (*Acer saccharum*) discoloured wood characteristics. Can J For Res 43:649–657

Barton CVM (2012) Advances in remote sensing of plant stress. Plant Soil 354:41–44

Battaglia M, Cherry ML, Beadle CL, Sands PJ, Hingston A (1998) Prediction of leaf area index in eucalypt plantations: effects of water stress and temperature. Tree Physiol 18:521–528

Battaglia M, Sands PJ (1998) Process-based forest productivity models and their application in forest management. For Ecol Manage 102:13–32

Baynes J (2004) Assessing forest canopy density in a highly variable landscape using Landsat data and FCD Mapper software. Aust For 67:247–253

Beadle CL (1997) Dynamics of leaf and canopy development. In: Nambiar EKS, Brown AG (eds) Management of soil, nutrients and water in tropical plantation forests. Australian Centre for International Agricultural Research Monograph No. 43, Canberra, pp 169–212

Behn G, McKinnell FH, Caccetta P, Vernes T (2001) Mapping forest cover, Kimberley Region of Western Australia. Aust For 64:80–87

Bengough AG, Castrignano A, Pagès L, van Noordwijk M (2000) Sampling strategies, scaling and statistics. In: Smit AM et al (eds) Root methods. Springer, Berlin, Heidelberg, New York, pp 147–173

Berger A, Gschwantner T, McRoberts RE, Schadauer K (2014) Effects of measurement errors on individual tree stem volume estimates for the Austrian National Forest Inventory. For Sci 60:14–24

Berninger F, Nikinmaa E (1998) Foliage area-sapwood area relationships of Scots pine (*Pinus sylvestris*) trees in different climates. Can J For Res 24:2263–2268

Berrill J-P, O'Hara KL (2014) Estimating site productivity in irregular stand structures by indexing the basal area or volume increment of the dominant species. Can J For Res 44:92–100

Bi H (1994) Volume equations for six *Eucalyptus* species on the south-east tablelands of New South Wales. State Forests NSW, Sydney, Research Report No. 23

Bi H (2000) Trigonometric variable-form taper equations for Australian eucalypts. For Sci 46:397–409

Bi H, Birk E, Turner J, Lambert M, Jurskis V (2001) Converting stem volume to biomass with additivity, bias correction, and confidence bands for two Australian tree species. N Z J For Sci 31:298–319

Bi H, Long Y (2001) Flexible taper equation for site-specific management of *Pinus radiata* in New South Wales, Australia. For Ecol Manage 148:79–91

Biging GS, Wensel LC (1988) The effect of eccentricity on the estimation of basal area and basal area increment of coniferous trees. For Sci 34:621–633

Bollandsås OM, Hanssen KH, Marthiniussen S, Næsset E (2008) Measures of spatial forest structure derived from airborne laser data are associated with natural regeneration patterns in an uneven-aged spruce forest. For Ecol Manage 255:953–961

Bollandsås OM, Næsset E (2007) Estimating percentile-based diameter distributions in uneven-sized Norway spruce stands using airborne laser scanner data. Scand J For Res 22:33–47

Bond BJ, Czarnomski NM, Cooper C, Day ME, Greenwood MS (2007) Developmental decline in height growth in Douglas-fir. Tree Physiol 27:441–453

Bontemps JD, Bouriaud O (2014) Predictive approaches to forest site productivity: recent trends, challenges and future perspectives. Forestry 87:109–128

Borden KA, Isaac ME, Thevathasan NV, Gordon AM, Thomas SC (2014) Estimating coarse root biomass with ground penetrating radar in a tree-based intercropping system. Agrofor Syst 88:657–669

Brandeis TJ, Delaney M, Parresol BR, Royer L (2006) Development of equations for predicting Puerto Rican subtropical dry forest biomass and volume. For Ecol Manage 233:133–142

Branthomme A, Saket M, Altrell D, Vuorinen P, Dalsgaard S, Andersson LGB (2004) National forest inventory manual, 4th edn. Forestry Department, Forest Resources Assessment Programme, Working paper 94/E. FAO, Rome

Brassard BW, Chen HYH, Bergeron Y, Paré D (2011) Coarse root biomass allometric equations for *Abies balsamea*, *Picea mariana*, *Pinus banksiana*, and *Populus tremuloides* in the boreal forest of Ontario, Canada. Biomass Bioenergy 35:4189–4196

Brazee NJ, Marra RE, Göcke L, Van Wassenaer P (2011) Non-destructive assessment of internal decay in three hardwood species of northeastern North America using sonic and electrical impedance tomography. Forestry 84:33–39

Breidenbach J, Gläser C, Schmidt M (2008) Estimation of diameter distributions by means of airborne laser scanner data. Can J For Res 38:1611–1620

Briggs DG, Kantavichai R, Turnblom EC (2008) Effects of precommercial thinning followed by a fertilization regime on branch diameter in coastal United States Douglas-fir plantations. Can J For Res 38:1564–1575

Brooks JR, Jiang L, Özçelik R (2008) Compatible stem volume and taper equations for Brutian pine, Cedar of Lebanon and Cilicica fir in Turkey. For Ecol Manage 256:147–151

Brooks JR, Wiant HV (2004) A simple technique for estimating cubic volume yields. For Ecol Manage 203:373–380

Brusa A, Bunker DE (2014) Increasing the precision of canopy closure estimates from hemispherical photography: blue channel analysis and under-exposure. Agric For Meteorol 195:102–107

Buckley TN, Roberts DW (2005) DESPOT, a process-based tree growth model that allocates carbon to maximize carbon gain. Tree Physiol 26:129–144

Burger JA (2009) Management effects on growth, production and sustainability of managed forest ecosystems: past trends and future directions. For Ecol Manage 258:2335–2346

Burrows WH, Hoffmann MB, Compton JF, Back PV (2001) Allometric relationships and community biomass stocks in white cypress pine (*Callitris glaucophylla*) and associated eucalypts of the Carnarvon area, south central Queensland. National Carbon Accounting System Technical Report No. 33. Australian Greenhouse Office, Canberra

Butnor JR, Doolittle JA, Johnsen KH, Samuelson L, Stokes T, Kress L (2003) Utility of ground-penetrating radar as a root biomass survey tool in forest systems. Soil Sci Soc Am J 67:1607–1615

Butnor JR, Pruyn ML, Shaw DC, Harmon ME, Mucciardi AN, Ryan MG (2009) Detecting defects in conifers with ground penetrating radar: applications and challenges. Forest Pathol 39:309–322

Cao L, Coops NC, Innes J, Dai JS, She GH (2014) Mapping above- and below-ground biomass components in subtropical forests using small-footprint LiDAR. Forests 5:1356–1373

Cao QV, Wang J (2011) Calibrating fixed- and mixed-effects taper equations. For Ecol Manage 262:671–673

Carvalho JP, Parresol BR (2003) Additivity in tree biomass components of Pyrenean oak (*Quercus pyrenaica* Willd.). For Ecol Manage 179:269–276

Case BS, Hall RJ (2008) Assessing prediction errors of generalized tree biomass and volume equations for the boreal forest region of west-central Canada. Can J For Res 38:878–889

Causton DR (1985) Biometrical, structural and physiological relationships among tree parts. In: Cannell MGR, Jackson JE (eds) Trees as crop plants. Institute of Terrestrial Ecology, Natural Environment Research Council, UK, pp 137–159

Cavanaugh KC, Gosnell JS, Davis SL, Ahumada J, Boundja P, Clark DB, Mugerwa B, Jansen PA, O'Brien TG, Rovero F, Sheil D, Vasquez R, Andelman S (2014) Carbon storage in tropical forests correlates with taxonomic diversity and functional dominance on a global scale. Global Ecol Biogeogr 23:563–573

Cescatti A (2007) Indirect estimates of canopy gap fraction based on the linear conversion of hemispherical photographs. Methodology and comparison with standard threshold techniques. Agric For Meteorol 143:1–12

Chambers JQS, Santos JD, Ribeiro RJ, Higuchi N (2001) Tree damage, allometric relationships, and aboveground net primary production in central Amazon Forest. For Ecol Manage 152:73–84

Chan SS, McCreight RW, Walstad JD, Spies TA (1986) Evaluating forest vegetative cover with computerized analysis of fisheye photographs. For Sci 32:1085–1091

Chapman L (2007) Potential applications of near infra-red hemispherical imagery in forest environments. Agric For Meteorol 143:151–156

Chasmer L, Hopkinson C, Treitz P (2006) Investigating laser pulse penetration through a conifer canopy by integrating airborne and terrestrial lidar. Can J Remote Sens 32:116–125

Chason JW, Baldocchi DD, Huston MA (1991) A comparison of direct and indirect methods for estimating forest canopy leaf area. Agric For Meteorol 57:107–128

Chave J et al (2005) Tree allometry and improved estimation of carbon stocks and balance in tropical forests. Oecologia 145:87–99

Chave J et al (2014) Improved allometric models to estimate the aboveground biomass of tropical trees. Global Change Biol. 20:3177–3190

Chave J, Condit R, Aguilar S, Hernandez A, Lao S, Perez R (2004) Error propagation and scaling for tropical forest biomass estimates. Philos Trans R Soc Lond B Biol Sci 359:409–420

Chave J, Coomes D, Jansen S, Lewis SL, Swenon S, Zanne AE (2009) Towards a worldwide wood economics spectrum. Ecol Lett 12:351–366

Chave J, Muller-Landau HC, Baker TR, Easdale TA, Ter Steege H, Webb CO (2006) Regional and phylogenetic variation of wood density across 2456 neotropical tree species. Ecol Appl 16:2356–2367

Cheng D-L, Niklas KJ (2007) Above- and below-ground biomass relationships across 1534 forested communities. Ann Bot 99:95–102

Chirici G, Barbati A, Maselli F (2007) Modelling of Italian forest net primary productivity by the integration of remotely sensed and GIS data. For Ecol Manage 246:285–295

Chojnacky DC, Heath LS, Jenkins JC (2014) Updated generalized biomass equations for North American tree species. Forestry 87:129–151

Cienciala E, Cerný M, Tatarinov J, Apltauer J, Exnerová Z (2006) Biomass functions applicable to Scots pine. Trees 20:483–495

Cienciala E, Exenerová Z, Schelhaas M-J (2008) Development of forest carbon stock and wood production in the Czech Republic until 2060. Ann For Sci 65:603–612

Cienciala E, Tatarinov FA (2006) Application of BIOME-BGC model to managed forests. 2. Comparison with long-term observations of stand production for major tree species. For Ecol Manage 237:252–266

Clark AC, Souter RA, Schlagel BE (1991) Stem profile equations for southern tree species. Research Paper SE-282. USDA Forest Service, Southeastern Forest Experiment Station, Asheville, NC

Coble DW, Wiant HV (2000) Centroid method: comparison of simple and complex proxy tree taper functions. For Sci 46:473–477

Cochran WG (1977) Sampling techniques, 3rd edn. Wiley, New York

Cohen R, Kaino J, Okello JA, Bosire JO, Kairo JG, Huxham M, Mencuccini M (2013) Propagating uncertainty to estimates of above-ground biomass for Kenyan mangroves: a scaling procedure from tree to landscape level. For Ecol Manage 310:968–982

Cole TG, Ewel JJ (2006) Allometric equations for four valuable tropical tree species. For Ecol Manage 229:351–360

Colgan MS, Swemmer T, Asner GP (2014) Structural relationships between form factor, wood density, and biomass in African savanna woodlands. Trees 28:91–102

Coll L, Potvin C, Messier C, Delagrange S (2008) Root architecture and allocation patterns of eight native tropical species with different successional status used in open-grown mixed plantations in Panama. Trees 22:585–596

Coops N (2002) Eucalypt forest structure and synthetic aperture radar backscatter: a theoretical analysis. Trees 16:28–46

Coops N, Bi H, Barnett P, Ryan P (1999) Estimating mean and current annual increments of stand volume in a regrowth eucalypt forest using historical landsat multi spectral scanner imagery. J Sust For 9:149–167

Coops N, Delahaye A, Pook E (1997) Estimation of forest leaf area index on the south coast of New South Wales using Landsat MSS data. Aust J Bot 45:757–769

Coops N, Stanford M, Old K, Dudzinski M, Stone C (2003a) Assessment of Dothistroma needle blight of *Pinus radiata* using airborne hyperspectral imagery. Phytopathology 93:1524–1532

Coops NC, Hilker T, Wulder MA, St-Onge B, Newnham G, Siggins A, Trofymow JA (2007) Estimating canopy structure of Douglas-fir forest stands from discrete-return LiDAR. Trees 21:295–319

Coops NC, Smith ML, Jacobsen KL, Martin M, Ollinger S (2004a) Estimation of plant and leaf area index using three techniques in a mature native eucalypt canopy. Austral Ecol 29:332–341

Coops NC, Stone C, Culvenor DS, Chisholm L (2004b) Assessment of crown condition in eucalypt vegetation by remotely sensed optical indices. J Environ Qual 33:956–964

Coops NC, Stone C, Culvenor DS, Chisholm LA, Merton RN (2003b) Chlorophyll content in eucalypt vegetation at the leaf and canopy scales as derived from high resolution spectral data. Tree Physiol 23:23–31

Coops NC, Waring RH, Brown SR, Running SW (2001) Comparisons of predictions of net primary production and seasonal patterns in water use derived with two forest growth models in Southwestern Oregon. Ecol Mod 142:61–81

Coops NC, Waring RH, Landsberg JJ (1998) Assessing forest productivity in Australia and New Zealand using a physiologically-based model driven with averaged monthly weather data and satellite imagery. For Ecol Manage 104:113–127

Coops NC, Wulder MA, Culvenor DS, St-Onge B (2004c) Comparison of forest attributes extracted from fine spatial resolution multispectral and lidar data. Can J Remote Sens 30:855–866

Cordero LDP, Kanninen M (2003) Provisional equations for estimating total and merchantable volume of *Tectona grandis* trees in Costa Rica. Forests Trees Livelihoods 13:345–349

Corona P, Chirici G, McRoberts RE, Winter S, Barbati A (2011) Contribution of large-scale forest inventories to biodiversity assessment and monitoring. For Ecol Manage 262:2061–2069

Coyle DR, Coleman MD, Aubrey DP (2008) Above- and below-ground biomass accumulation, production, and distribution of sweetgum and loblolly pine grown with irrigation and fertilization. Can J For Res 38:1335–1348

Cramer MD (2012) Unravelling the limits to tree height: a major role for water and nutrient trade-offs. Oecologia 169:61–72

Crous JW, Morris AR, Scholes MC (2009) Effects of phosphorus and potassium fertiliser on stem form, basic wood density and stem nutrient content of *Pinus patula* at various stem heights. Aust For 72:99–111

Cunia T (1965) Some theory on reliability of volume estimates in a forest inventory sample. For Sci 11:115–128

da Costa Freitas C, de Souza Soler L, Sant'Anna SJS, Dutra LV, dos Santos JR, Mura JC, Correia AH (2008) Land use and land cover mapping in the Brazilian Amazon using polarimetric airborne P-band SAR data. IEEE Trans Geosci Rem Sens 46:2956–2970

Danjon F, Reubens B (2008) Assessing and analyzing 3D architecture of woody root systems, a review of methods and applications in tree and soil stability, resource acquisition and allocation. Plant Soil 303:1–34

Danson FM, Hetherington D, Monsdorf F, Koetz B, Allgower B (2007) Forest canopy gap fraction from terrestrial laser scanning. IEEE Geosci Remote Sens Lett 4:157–160

Davi H, Baret F, Huc R, Dufrêne E (2008) Effect of thinning on LAI variance in heterogeneous forest. For Ecol Manage 256:890–899

Davis LS, Johnson KN, Bettinger P, Howard EH (2001) Forest management to sustain ecological, economic, and social values, 4th edn. Waveland Press, Long Grove, IL

De Grandi DG, Lucas RM, Kropacek J (2009) Analysis by wavelet frames of spatial statistics in SAR data for characterizing structural properties of forests. IEEE Trans Geosci Remote Sens 47:494–507

Dean C (2003) Calculation of wood volume and stem taper using terrestrial single-image close-range photogrammetry and contemporary software tools. Silva Fenn 37:359–380

Dean C, Roxburgh S (2006) Improving visualisation of mature, high-carbon-sequestering forests. Forest Biometry Model Inf Sci 1:48–69

Dean TJ (2004) Basal area increment and growth efficiency as functions of canopy dynamics and stem mechanics. For Sci 50:106–116

Dean TJ, Roberts SD, Gilmore DW, Maguire DA, Long JN, O'Hara KL, Seymour RS (2002) An evaluation of the uniform stress hypothesis based on stem geometry in selected North American conifers. Trees 16:559–568

Deleuze C, Houllier F (2002) A flexible radial increment taper equation derived from a process-based carbon partitioning model. Ann For Sci 59:141–154

Demarez V, Duthoit S, Baret F, Weiss M, Dedieu G (2008) Estimation of leaf area and clumping indexes of crops with hemispherical photographs. Agric For Meteorol 148:644–655

Deraedt W, Ceulemans R (1998) Clonal variability in biomass production and conversion efficiency of poplar during the establishment of a short rotation coppice plantation. Biomass Bioenergy 15:391–398

Di Iorio A, Lassere B, Scippa GS, Chiatante D (2005) Root system architecture of *Quercus pubescens* trees growing on different sloping conditions. Ann Bot 95:351–361

Diéguez-Aranda U, Burkhart HE, Amateis RL (2006a) Dynamic site model for loblolly pine (*Pinus taeda* L.) plantations in the United States. For Sci 52:262–272

Diéguez-Aranda U, Castedo-Dorado F, Álvarez-González JG, Rojo A (2006b) Compatible taper functions for Scots pine plantations in northwest Spain. Can J For Res 36:1190–1205

Dieters M, Brawner J (2007) Productivity of *Pinus elliottii*, *P. caribaea* and their F_1 and F_2 hybrids to 15 years in Queensland, Australia. Ann For Sci 64:691–698

Dillen SY, Marron N, Bastien C, Riciotti L, Salani F, Sabatti M, Pinel PC, Rae AM, Taylor G, Ceulemans R (2007) Effects of environment and progeny on biomass estimations of five hybrid poplar families grown at three contrasting sites across Europe. For Ecol Manage 252:12–23

Ditzer T, Glauner R, Förster M, Köhler P, Huth A (2000) The process-based stand growth model Formix 3-Q applied in a GIS environment for growth and yield analysis in a tropical rainforest. Tree Physiol 20:367–381

Djomo AN, Ibrahima A, Saborowski J, Gravenhorst G (2010) Allometric equations for biomass estimations in Cameroon and pan moist tropical equations including biomass data from Africa. For Ecol Manage 260:1873–1885

Djomo AN, Knohl A, Gravenhorst G (2011) Estimations of total ecosystem carbon pools distribution and carbon biomass current annual increment of a moist tropical forest. For Ecol Manage 261:1448–1459

Domec JC, Lachenbruch B, Meinzer FC, Woodruff DR, Warren JM, McCulloh KA (2008) Maximum height in a conifer is associated with conflicting requirements for xylem design. Proc Natl Acad Sci U S A 105:12069–12074

Donoghue DNM, Watt PJ, Cox NJ, Dunford RW, Wilson J, Stables S, Smith S (2004) An evaluation of the use of satellite data for monitoring early development of young Sitka spruce plantation forest growth. Forestry 77:383–396

Dovey SB, du Toit B (2006) Calibration of the LAI-2000 canopy analyser with leaf area index in a young eucalypt stand. Trees 20:273–277

Drake JE, Davis SC, Raetz LM, DeLucia EH (2011) Mechanisms of age-related changes in forest production: the influence of physiological and successional changes. Global Change Biol 17:1522–1535

Drake JE, Raetz LM, Davis SC, DeLucia EH (2010) Hydraulic limitation not declining nitrogen availability causes the age-related photosynthetic decline in loblolly pine (*Pinus taeda* L.). Plant Cell Environ 33:1756–1766

Draper NR, Smith H (1988) Applied regression analysis, 3rd edn. Wiley, New York

Drezet PML, Quegan S (2007) Satellite-based radar mapping of British forest age and net ecosystem exchange using ERS tandem coherence. For Ecol Manage 238:65–80

Droppelmann KJ, Berliner PR (2000) Biometric relationships and growth of pruned and non-pruned *Acacia saligna* under runoff irrigation in northern Kenya. For Ecol Manage 126:349–359

Du L, Zhou T, Zou ZH, Zhao X, Huang KC, Wu H (2014) Mapping forest biomass using remote sensing and national forest inventory in China. Forests 5:1267–1283

Ducey MJ, Gove JH, Valentine HT (2004) A walkthrough solution to the boundary overlap problem. For Sci 50:427–435

Ducey MJ, Williams MS (2011) Comparison of Hossfeld's method and two modern methods for volume estimation of standing trees. West J Appl For 26:19–23

Dutilleul P, Han L, Smith DL (2008) Plant light interception can be explained via computed tomography scanning: demonstration with pyramidal cedar (*Thuja occidentalis*, Fastigiata). Ann Bot 101:19–23

Eamus D, McGuinness K, Burrows W (2000) Review of allometric relationships for estimating woody biomass for Queensland, the Northern Territory and Western Australia. National Carbon Accounting System Technical Report No. 5a. Australian Greenhouse Office, Canberra

Eckstein D, Dujesiefken D (1999) Long-term effects in trees due to increment borings. Dendrochronologia 16–17:205–211

Falkowski MJ, Smith AMS, Hudak AT, Gessler PE, Vierling LA, Crookston NL (2006) Automated estimation of individual conifer tree height and crown diameter via two-dimensional spatial wavelet analysis of lidar data. Can J Remote Sens 32:153–161

Fang Z, Bailey RL (1999) Compatible volume and taper models with coefficients for tropical species on Hainan Island in southern China. For Sci 45:85–100

Fassnacht K, Gower ST, Norman JM, McMurtrie RE (1994) A comparison of optical and direct methods for estimating foliage surface area index in forests. Agric For Meteorol 71:183–207

Feldpausch TR et al (2012) Tree height integrated into pantropical forest biomass estimates. Biogeosciences 9:3381–3403

Fensham RJ, Fairfax RJ (2002) Aerial photography for assessing vegetation change: a review of applications and the relevance of findings for Australian vegetation history. Aust J Bot 50:415–429

Fensham RJ, Fairfax RJ, Holman JE, Whitehead PJ (2002) Quantitative assessment of vegetational structural attributes from aerial photography. Int J Remote Sens 23:2293–2317

Fernández ME, Gyenge J (2009) Testing Binkley's hypothesis about the interaction of individual tree water use efficiency and growth efficiency with dominance patterns in open and close canopy stands. For Ecol Manage 257:1859–1865

Figueiredo A, Machado SA, Carneiro MRA (2000) Testing accuracy of log volume calculation procedures against water displacement techniques (xylometer). Can J For Res 30:990–997

Filho AF, Schaaf LB (1999) Comparison between predicted volumes estimated by taper equations and true volumes obtained by the water displacement technique (xylometer). Can J For Res 29:451–461

Fleck S, Niinemets Ü, Cescatti A, Tenhunen JD (2003) Three-dimensional lamina architecture alters light-harvesting efficiency in *Fagus*: a leaf-scale analysis. Tree Physiol 23:577–589

Flewelling JW, Strunk JL (2013) The walk through and fro estimator for edge bias avoidance. For Sci 59:223–230

Fonweban J, Gardiner B, Auty D (2012) Variable-top merchantable volume equations for Scots pine (*Pinus sylvestris*) and Sitka spruce (*Picea sitchensis* (Bong.) Carr.) in Northern Britain. Forestry 85:237–253

Fonweban J, Gardiner B, Macdonald E, Auty D (2011) Taper functions for Scots pine (*Pinus sylvestris* L.) and Sitka spruce (*Picea sitchensis* (Bong.) Carr.) in Northern Britain. Forestry 84:49–60

Fonweban JN (1997) Effect of log formula, log length and method of measurement on the accuracy of volume estimates for three tropical timber species in Cameroon. Comm For Rev 76:114–120

Forslund RR (1982) A geometrical tree volume model based on the location of the centre of gravity of the bole. Can J For Res 12:215–221

Fourcaud T, Blaise F, Lac P, Castéra P, de Reffye P (2003) Numerical modelling of shape regulation and growth stresses in trees. II. Implementation in the AMAPpara software and simulation of tree growth. Trees 17:31–39

Fourcaud T, Lac P (2003) Numerical modelling of shape regulation and growth stresses in trees. 1. An incremental static finite element formulation. Trees 17:23–30

Fourcaud T, Zhang X, Stokes A, Lambers H, Körner C (2008) Plant growth modelling and applications: the increasing importance of plant architecture in growth models. Ann Bot 101:1053–1063

Fournier RA, Mailly D, Walter J-MN, Soudani K (2003) Indirect measurement of forestry canopy structure from in situ optical sensors. In: Wulder MA, Franklin SE (eds) Remote sensing of forest environments: concepts and case studies. Kluwer, Dordrecht, pp 77–113

Freedman B (1984) The relationship between the above-ground dry weight and diameter for a wide size range of erect land plants. Can J For Res 62:2370–2374

Frescino TS, Edwards TC, Moisen GG (2001) Modeling spatially explicit forest structural attributes using generalized additive models. J Veg Sci 12:15–26

Freund RJ, Wilson WJ, Sa P (2006) Regression analysis, 2nd edn. Academic Press, London

Fujisaki I, Evans DL, Moorhead RJ, Irby DW, Mohammadi-Aragh MJ, Roberts SD, Gerard PD (2008) Stand assessment through lidar-based forest visualization using immersive virtual environment technology. For Sci 54:1–7

Furnival GM, Valentine HT, Gregoire TG (1986) Estimation of log volume by importance sampling. For Sci 32:1073–1078

Furst C, Vacik H, Lorz C, Makeschin F, Podrazky V (2007) Meeting the challenges of process-oriented forest management. For Ecol Manage 248:1–2

Garber SM, Maguire DA (2003) Modeling stem taper of three central Oregon species using nonlinear mixed effects models and autoregressive error structures. For Ecol Manage 179:507–522

Gaul D, Hertel D, Borken W, Matzner E, Leuschner C (2008) Effects of experimental drought on the fine root system of mature Norway spruce. For Ecol Manage 256:1151–1159

Gertner GZ (1990) The sensitivity of error in stand volume estimation. Can J For Res 20:800–804

Gholz HL, Vogel SA, Cropper WP, McKelvey K, Ewel KC, Teskey RO, Curran PJ (1991) Dynamics of canopy structure and light interception in Pinus elliottii stands, north Florida. Ecol Monogr 61:33–51

Gifford RM (2000) Carbon contents of above-ground tissues of forest and woodland trees. National Carbon Accounting System Technical Report No. 22. Australian Greenhouse Office, Canberra

Gilmore DW, Seymour RS (2004) Foliage-sapwood area equations for balsam fir require local validation. For Sci 50:566–570

Gobakken T, Næsset E (2005) Weibull and percentile methods for lidar-based estimation of basal area distribution. Scand J For Res 20:490–502

Gobakken T, Næsset E (2008) Assessing effects of laser point density, ground sampling intensity, and field sample plot size on biophysical stand properties derived from airborne laser scanner data. Can J For Res 38:1095–1109

Goetz SJ, Prince SD, Goward SN, Thawley MM, Small J (1999) Satellite remote sensing of primary production: an improved production efficiency modeling approach. Ecol Mod 122:239–255

Goméz-García E, Crecente-Campo F, Diéguez-Aranda U (2013) Selection of mixed-effects parameters in a variable-exponent taper equation for birch trees in northwestern Spain. Ann For Sci 70:707–715

Goodman RC, Phillips OL, Baker TR (2014) The importance of crown dimensions to improve tropical tree biomass estimates. Ecol Appl 24:680–698

Goodwin A (2009) A cubic tree taper model. Aust For 72:87–98

Goodwin N, Turner R, Merton R (2005) Classifying *Eucalyptus* forests with high spatial and spectral resolution imagery: an investigation of individual species and vegetation communities. Aust J Bot 53:337–345

Gouveia A, Freitas H (2008) Intraspecific competition and water use efficiency in *Quercus suber*: evidence of an optimum tree density? Trees 22:521–530

Gräfstrom A, Ringvall H (2013) Improving forest field inventories by using remote sensing data in novel sampling designs. Can J For Res 43:1015–1022

Gregoire TG, Ståhl G, Næsset E, Gobakken T, Nelson R, Holm S (2011) Model-assisted estimator of biomass in a LIDAR sample survey in Hedmark County, Norway. Can J For Res 41:83–95

Gregoire TG, Valentine HT (2008) Sampling strategies for natural resources and the environment. Chapman & Hall/CRC, Boca Raton, FL

Gregoire TG, Valentine HT, Furnival GM (1986) Estimation of bole volume by importance sampling. Can J For Res 16:554–557

Gregoire TG, Valentine HT, Furnival GM (1995) Sampling methods to estimate foliage and other characteristics of individual trees. Ecology 76:1181–1194

Gregoire TG, Zedaker SM, Nicholas NS (1990) Modeling relative error in stem basal area estimates. Can J For Res 20:496–502

Grierson PF, Adams MA, Attiwill PM (1992) Estimates of carbon storage in the above-ground biomass of Victoria's forests. Aust J Bot 40:631–640

Grissino-Mayer HD (2003) A manual and tutorial for the proper use of an increment borer. Tree-Ring Res 59:63–79

Grote R, Reiter IM (2004) Competition-dependent modelling of foliage biomass in forest stands. Trees 18:596–607

Guo L, Chen J, Cui X, Fan B, Lin H (2013a) Application of ground penetrating radar for coarse root detection and quantification: a review. Plant Soil 362:1–23

Guo L, Lin H, Fan BH, Cui XH, Chen J (2013b) Forward simulation of root's ground penetrating radar signal: simulator development and validation. Plant Soil 372:487–505

Hackett C, Vanclay JK (1998) Mobilizing expert knowledge of tree growth with the PLANTGRO and INFER systems. Ecol Mod 106:233–246

Hall RJ (2003) The roles of aerial photographs in forestry remote sensing image analysis. In: Wulder MA, Franklin SE (eds) Remote sensing of forest environments: concepts and case studies. Kluwer, Dordrecht, pp 47–75

Hall RJ, Skakun RS, Arsenault EJ, Case BS (2006) Modeling forest stand structure attributes using Landsat ETM+ data: application to mapping of aboveground biomass and stand volume. For Ecol Manage 225:378–390

Hall SA, Burke IC, Box DO, Kaufmann MR, Stoker JM (2005) Estimating stand structure using discrete-return lidar: an example from low density, fire prone ponderosa pine forests. For Ecol Manage 208:189–209

Hamilton F, Brack C (1999) Stand volume estimates from modelling inventory data. Aust For 62:360–367

Hamilton SD, Brodie G, O'Dwyer C (2005) Allometric relationships for estimating biomass in grey box (*Eucalyptus microcarpa*). Aust For 68:267–273

Hanssen KH, Solberg S (2007) Assessment of defoliation during a pine sawfly outbreak: calibration of airborne laser scanning data with hemispherical photography. For Ecol Manage 250:9–16

Hapca AH, Mothe F, Leban J-M (2007) A digital photographic method for 3D reconstruction of standing tree shape. Ann For Sci 64:631–637

Harcombe PA, Greene SE, Kramer MG, Acker SA, Spies TA, Valentine T (2004) The influence of fire and windthrow dynamics on a coastal spruce-hemlock forest in Oregon, USA, based on aerial photographs spanning 40 years. For Ecol Manage 194:71–82

Hayashi R, Weiskittel A, Sader S (2014) Assessing the feasibility of low-density LiDAR for stand inventory attribute predictions in complex and managed forests of northern Maine, USA. Forests 5:363–383

Hayward WJ (1987) Volume and taper of *Eucalyptus regnans* grown in the central north island of New Zealand. N Z J For Sci 17:109–120

Helmisaari H-S, Dermoe J, Nöjd P, Kukkola M (2007) Fine root biomass in relation to site and stand characteristics in Norway spruce and Scots pine stands. Tree Physiol 27:1493–1504

Henning JG, Radtke PJ (2006) Detailed stem measurement of standing trees from ground-based scanning Lidar. For Sci 52:67–80

Henry M, Besnard A, Asante WA, Eshun J, Adu-Bredu S, Valentini R, Bernoux M, Saint-André L (2010) Wood density, phytomass variations within and among trees, and allometric equations in a tropical rainforest of Africa. For Ecol Manage 260:1375–1388

Henry M, Picard N, Trotta C, Manlay RJ, Valentini R, Bernoux M, Saint-André L (2011) Estimating tree biomass of sub-Saharan African forests: a review of available allometric equations. Silva Fenn 45:477–569

Henskens FL, Battaglia M, Cherry ML, Beadle CL (2001) Physiological basis of spacing effects on tree growth and form in *Eucalyptus globulus*. Trees 15:365–377

Herrero C, Juez L, Tejedor C, Pando V, Bravo F (2014) Importance of root system in total biomass for *Eucalyptus globulus* in northern Spain. Biomass Bioenergy 67:212–222

Heurich M (2008) Automatic recognition and measurement of single trees based on data from airborne laser scanning over the richly structured national forests of the Bavarian Forest National Park. For Ecol Manage 255:2416–2433

Hilker T, Wulder MA, Coops NC (2008) Update of forest inventory data with lidar and high spatial resolution satellite imagery. Can J Remote Sens 34:5–12

Hill A, Breschan J, Mandallaz D (2014) Accuracy assessment of timber volume maps using forest inventory data and LiDAR canopy height models. Forests 5:2253–2275

Hjelm B, Johansson T (2012) Volume equations for poplars growing on farmland in Sweden. Scand J For Res 27:561–566

Holdaway RJ, Allen RB, Clinton PW, Davis MR, Coomes DA (2008) Intraspecific changes in forest canopy allometries during self-thinning. Funct Ecol 22:460–469

Holdaway RJ, McNeill SJ, Mason NWH, Carswell FE (2014) Propagating uncertainty in plot-based estimates of forest carbon stock and carbon stock change. Ecosystems 17:627–640

Holmgren J, Nilsson M, Olsson H (2003) Estimation of tree height and stem volume using airborne laser scanning. For Sci 49:419–428

Holopainen M, Vastaranta M, Hyyppä J (2014) Outlook for the next generation's precision forestry in Finland. Forests 5:1682–1694

Hopkinson C, Chasmer L, Young-Pow C, Treitz P (2004) Assessing forest metrics with a ground-based scanning lidar. Can J For Res 34:575–583

Huang S (1997) Development of compatible height and site index models for young and mature stands within an ecosystem-based management framework. In: Amaro A, Tomé M (eds) Empirical and process-based models for forest tree and stand growth simulation. Edições Salamandra, Lisboa, pp 61–98

Hultnas M, Nylinder M, Ågren A (2013) Predicting the green density as a means to achieve the volume of Norway spruce. Scand J For Res 28:257–265

Husch B, Beers TW, Kershaw JA (2003) Forest mensuration, 4th edn. Wiley, NJ

Hutchison J, Manica A, Swetnam R, Balmford A, Spalding M (2014) Predicting global patterns in mangrove forest biomass. Conserv Lett 7:233–240

Hyyppä J, Holopainen M, Olsson H (2012a) Laser scanning in forests. Remote Sens 4:2919–2922

Hyyppä J, Yu XW, Hyyppä H, Vastaranta M, Holopainen M, Kukko A, Kaartinen H, Jaakkola A, Vaja M, Koskinen J, Alho P (2012b) Advances in forest inventory using airborne laser scanning. Remote Sens 4:1190–1207

Ikonen V-P, Kellomäki S, Väisänen H, Peltola H (2006) Modelling the distribution of diameter growth along the stem in Scots pine. Trees 20:391–402

Iles K (2003) A sampler of inventory topics. Kim Iles & Associates, Nanaimo, BC

Jacobs MR (1954) The effect of wind sway on the form and development of *Pinus radiata* D. Don. Aust J Bot 2:35–51

Jalkanen A, Mäkipää R, Ståhl G, Lehtonen A, Petersson H (2005) Estimation of the biomass stock of trees in Sweden: comparison of biomass equations and age-dependent biomass expansion factors. Ann For Sci 62:845–851

Jenkins JC, Chojnacky DC, Heath LS, Birdsey RA (2003) National-scale biomass estimators for United States tree species. For Sci 49:12–35

Jiang H, Apps MJ, Zhang Y, Peng C, Woodward PM (1999) Modelling the spatial pattern of net primary productivity in Chinese forests. Ecol Mod 122:275–288

Jiang L, Brooks JR, Wang J (2005) Compatible taper and volume equations for yellow-poplar in West Virginia. For Ecol Manage 213:399–409

Johansson T (2007) Biomass production and allometric above- and below-ground relations for young birch stands planted at four spacings on abandoned farmland. Forestry 80:41–52

Jonckheere I, Fleck S, Nackaerts K, Muys B, Coppin P, Weiss M, Baret F (2004) Review of methods for in situ leaf area determination. Part I. Theories, sensors and hemispherical photography. Agric For Meteorol 121:19–35

Jonckheere I, Muys B, Coppin P (2005a) Allometry and evaluation of *in situ* optical LAI determination in Scots pine: a case study in Belgium. Tree Physiol 25:723–732

Jonckheere I, Muys B, Coppin P (2005b) Derivative analysis for *in situ* high-dynamic range hemispherical photography and its application in forest stands. IEEE Geosci Remote Sens Lett 2:236–300

Jonckheere I, Nackaerts K, Muys B, Coppin P (2005c) Assessment of automatic gap fraction estimation of forests from digital hemispherical photography. Agric For Meteorol 132:96–114

Jourdan C, Silva EV, Gonçalves JLM, Ranger J, Laclau J-P (2008) Fine root production and turnover in Brazilian *Eucalyptus* plantations under contrasting nitrogen fertilization regimes. For Ecol Manage 256:396–404

Jupp DLB, Culvenor DS, Lovell JL, Newnham GJ, Strahler AH, Woodcock CE (2008) Estimating forest LAI profiles and structural parameters using a ground-based laser called 'Echidna®'. Tree Physiol 29:171–181

Kajimoto T, Matsuura Y, Osawa A, Abaimov AP, Zryanova OA, Isaev AP, Yefremov DP, Mori S, Koike T (2006) Size-mass allometry and biomass allocation of two larch species growing on the continuous permafrost region in Siberia. For Ecol Manage 222:314–325

Kalliovirta J, Laasasenaho J, Kangas A (2005) Evaluation of the laser-relascope. For Ecol Manage 204:181–194

Kangas A (2006) Mensurational aspects. In: Kangas A, Maltamo M (eds) Forest inventory methodology and applications. Springer, Dordrecht, pp 53–63

Kangas A, Maltamo M (eds) (2006) Forest inventory methodology and applications. Springer, Dordrecht

Kankare V, Joensuu M, Vauhkonen J, Holopainen M, Tanhuanpää T, Vastaranta M, Hyyppä J, Hyyppä H, Alho P, Rikala J, Sipi M (2014) Estimation of the timber quality of Scots pine with terrestrial laser scanning. Forests 5:1879–1895

Keane RE, Reinhardt ED, Scott J, Gray K, Reardon J (2005) Estimating forest canopy bulk density using six indirect methods. Can J For Res 35:725–739

Kenzo T, Furutani R, Hattori D, Kendawang JJ, Tanaka S, Sakurai K, Ninomiya I (2009a) Allometric equations for accurate estimation of above-ground biomass in logged-over tropical rainforests in Sarawak, Malaysia. J For Res 14:365–372

Kenzo T, Ichie T, Hattori D, Itioka T, Handa C, Ohkubo T, Kendawang JJ, Nakamura M, Sakaguchi M, Takahashi N, Okamoto M, Tanaka-Oda A, Ninomiya I (2009b) Development of allometric relationships for accurate estimation of above- and below-ground biomass in tropical secondary forests in Sarawak, Malaysia. J Trop Ecol 25:371–386

Khan MNI, Faruque O (2010) Allometric relationships for predicting the stem volume in a *Dalbergia sissoo* Roxb. plantation in Bangladesh. iForest 3:153–158

Khan MNI, Suwa R, Hagihara A (2005) Allometric relationships for estimating the above ground phytomass and leaf area of mangrove *Kandelia candel* (L.) Druce trees in the Manko Wetland, Okinawa Island, Japan. Trees 19:266–272

Kimsey MJ, Moore J, McDaniel P (2008) A geographically weighted regression analysis of Douglas-fir site index in north central Idaho. For Sci 54:356–366

Kinerson RS, Ralston CW, Wells CG (1977) Carbon cycling in a loblolly pine plantation. Oecologia 29:1–10

Kleinn C, Vilčko F (2006a) Design-unbiased estimation for point-to-tree distance sampling. Can J For Res 36:1407–1414

Kleinn C, Vilčko F (2006b) A new empirical approach for estimation in k-tree sampling. For Ecol Manage 237:522–533

Köhl M, Kushwaha SPS (1994) A four-phase sampling method for assessing stand volume using Landsat-TM-data, aerial photography and field assessments. Comm For Rev 73:35–42

Koike F (1985) Reconstruction of two-dimensional tree and forest canopy profiles using photographs. J Appl Ecol 22:921–929

Komiyama A, Ong JE, Poungparn S (2008) Allometry, biomass, and productivity of mangrove forests: a review. Acquat Bot 89:128–137

Koskela L, Nummi T, Wenzel S, Kivinen V-P (2006) On the analysis of cubic smoothing spline-based stem curve prediction for forest harvesters. Can J For Res 36:2909–2920

Kumar VSK, Tewari VP (1999) Above-ground biomass tables for *Azadirachta indica* a. Juss. Int For Rev 1:109–111

Labrecque S, Fournier RA, Luther JE, Piercy D (2006) A comparison of four methods to map biomass from Landsat-TM and inventory data in western Newfoundland. For Ecol Manage 226:129–144

Lai JS, Yang B, Lin DM, Kerkhoff AJ, Ma KP (2013) The allometry of coarse root biomass: log-transformed linear regression or nonlinear regression? PLoS One 8:219–225

Landsberg J, Sands P (2011) Physiological ecology of forest production. Academic Press, London

Landsberg JJ, Waring RH, Coops NC (2003) Performance of the forest productivity model 3-PG applied to a wide range of forest types. For Ecol Manage 172:199–214

Lang ARG, McMurtrie RE (1992) Total leaf area of single trees of *Eucalyptus grandis* and *Pinus radiata* estimated from transmittances of the sun's beam. Agric For Meteorol 58:79–92

Lang ARG, McMurtrie RE, Benson ML (1991) Validity of surface area indices of *Pinus radiata* estimated from transmittance of the sun's beam. Agric For Meteorol 57:157–170

Lappi J (2006) A multivariate, nonparametric stem-curve prediction method. Can J For Res 36:1017–1027

Lasserre B, Chirici G, Chiavetta U, Garfi V, Tognetti R, Drigo R, DiMartino P, Marchetti M (2011) Assessment of potential bioenergy from coppice forests through the integration of remote sensing and field surveys. Biomass Bioenergy 35:716–724

Latta G, Temesgen H, Barrett TM (2009) Mapping and imputing potential productivity of Pacific Northwest forests using climate variables. Can J For Res 39:1197–1207

Lavigne MB, Krasowski MJ (2007) Estimating coarse root biomass of balsam fir. Can J For Res 37:991–998

Le Maire G, Davi H, Soudani K, François C, Le Dantec V, Dufrêne E (2005) Modeling annual production and carbon fluxes of a large managed temperate forest using forest inventories, satellite data and field measurements. Tree Physiol 25:859–872

Le Roux X, Lacointe A, Escobar-Gutiérrez A, Le Dizès S (2001) Carbon-based models of individual tree growth: a critical appraisal. Ann For Sci 58:469–506

Lee EH, Tingey DT, Beedlow PA, Johnson MG, McKane RB (2004) A spatial analysis of fine-root biomass from stand data in the Pacific Northwest. Can J For Res 34:2169–2180

Lefsky MA (2010) A global forest canopy height map from the Moderate Resolution Imaging Spectroradiometer and the Geoscience Laser Altimeter System. Geophys Res Lett 37:15

Legner N, Fleck S, Leuschner C (2013) Low light acclimation in five temperate broad-leaved tree species of different successional status: the significance of a shade canopy. Ann For Sci 70:557–570

Lehtonen A, Cienciala E, Tatarinov F, Mäkipää R (2007) Uncertainty estimation of biomass expansion factors for Norway spruce in the Czech Republic. Ann For Sci 64:133–140

Lehtonen A, Mäkipää R, Heikkinen J, Sievänen R, Liski J (2004) Biomass expansion factors (BEFs) for Scots pine, Norway spruce and birch according to stand age for boreal forests. For Ecol Manage 188:211–224

Leites LP, Zubizarreta-Gerendiain A, Robinson AP (2013) Modeling mensurational relationships of plantation-grown loblolly pine (*Pinus taeda* L.) in Uruguay. For Ecol Manage 289:455–462

Levia DF (2008) A generalized allometric equation to predict foliar dry weight on the basis of trunk diameter for eastern white pine (*Pinus strobus* L.). For Ecol Manage 255:1789–1792

Li RX, Weiskittel A, Dick AR, Kershaw JA, Seymour RS (2012) Regional stem taper equations for eleven conifer species in the Acadian region of North America: development and assessment. North J Appl For 29:5–14

Li RX, Weiskittel AR (2010) Comparison of model forms for estimating stem taper and volume in the primary conifer species of the North American Acadian Region. Ann For Sci 67:302

Li Z, Kurz WA, Apps MJ, Beukema SJ (2003) Belowground biomass dynamics in the Carbon Budget Model of the Canadian Forest Sector: recent improvements and implications for the estimation of NPP and NEP. Can J For Res 33:126–136

Liang XL, Kankare V, Yu XW, Hyyppä J, Holopainen M (2014) Automated stem curve measurement using terrestrial laser scanning. IEEE Trans Geosci Remote Sens 52:1739–1748

Liddell MJ, Nieullet N, Campoe OC, Freiberg M (2007) Assessing the above-ground biomass of a complex tropical rainforest using a canopy crane. Austral Ecol 32:43–58

Lisa G, Faber-Langendoen D (2007) Development of stand structural stage indices to characterize forest conditions in upstate New York. For Ecol Manage 249:158–170

Louw JH, Scholes MC (2006) Site index functions using site descriptors for *Pinus patula* plantations in South Africa. For Ecol Manage 225:94–103

Lovell JL, Jupp DLB, Culvenor DS, Coops NC (2003) Using airborne and ground-based ranging lidar to measure canopy structure in Australian forests. Can J Remote Sens 29:607–662

Lovell JL, Jupp DLB, Newnham GJ, Coops NC, Culvenor DS (2005) Simulation study for finding optimal lidar acquisition parameters for forest height retrieval. For Ecol Manage 214:398–412

Lu D (2005) Integration of vegetation inventory data and Landsat TM image for vegetation classification in the western Brazilian Amazon. For Ecol Manage 213:369–383

Lusk CH, Warton DI (2007) Global meta-analysis shows that relationships of leaf mass per area with species shade tolerance depend on leaf habit and ontogeny. New Phytol 176:764–774

Macfarlane C, Arndt SK, Livesley SJ, Edgar AC, White DA, Adams MA, Eamus D (2007a) Estimation of leaf area index in eucalypt forest with vertical foliage, using cover and fullframe fisheye photography. For Ecol Manage 242:756–773

Macfarlane C, Hoffman M, Eamus D, Kerp N, Higginson S, McMurtrie R, Adams M (2007b) Estimation of leaf area index in eucalypt forest using digital photography. Agric For Meteorol 143:176–188

Macfarlane C, Ryu Y, Ogden GN, Sonnentag O (2014) Digital canopy photography: exposed and in the raw. Agric For Meteorol 197:244–253

MacFarlane DW, Green EJ, Burkhart HE (2000) Population density influences assessment and application of site index. Can J For Res 30:1472–1475

Magnusson M, Fransson ES (2005) Estimation of forest stem volume using multispectral optical satellite and tree height data in combination. Scand J For Res 20:431–440

Magnusson M, Fransson ES, Holmgren J (2007) Effects on estimation accuracy of forest variables using different pulse density of laser data. For Sci 53:619–626

Magnusson S, Picard N, Kleinn C (2008) A gamma-Poisson distribution of point to *k* nearest event distance. For Sci 54:429–441

Mäkelä A (2002) Derivation of stem taper from pipe theory in a carbon balance framework. Tree Physiol 22:891–905

Mäkelä A, Landsberg J, Ek AR, Burk TE, Ter-Mikaelian M, Ågren GI, Oliver CD, Puttonen P (2000) Process-based models for ecosystem management: current state of the art and challenges for practical implementation. Tree Physiol 20:289–298

Mäkelä A, Vanninen P (2000) Estimation of fine root mortality and growth from simple measurements: a method based on system dynamics. Trees 14:316–323

Malimbwi RE, Philip MS (1989) A compatible taper/volume estimation system for *Pinus patula* at Sao Hill forest project, Southern Tanzania. For Ecol Manage 27:109–115

Mallinis G, Koutsias N, Makras A, Karteris M (2004) Forest parameters estimation in a European Mediterranean landscape using remotely sensed data. For Sci 50:450–460

Maltamo M, Eerikäinen K, Packalén P, Hyyppä J (2006a) Estimation of stem volume using laser-scanning based canopy height metrics. Forestry 79:17–29

Maltamo M, Hyyppä J, Malinen J (2006b) A comparative study of the use of laser scanner data and field measurements in the prediction of crown height in boreal forests. Scand J For Res 21:231–238

Maltamo M, Korhonen KT, Packalén P, Mehtätalo L, Suvanto A (2007a) Testing the usability of angle count sample plots as ground truth in airborne laser scanning-based forest inventories. Forestry 80:73–81

Maltamo M, Malinen J, Packalén P, Suvanto A, Kangas J (2006c) Nonparametric estimation of stem volume using airborne laser scanning, aerial photography, and stand-register data. Can J For Res 36:426–436

Maltamo M, Mustonen K, Hyyppä J, Pitkänen J, Yu X (2004) The accuracy of estimating individual tree variables with airborne laser scanning in a boreal nature reserve. Can J For Res 34:1791–1801

Maltamo M, Packalén P, Yu X, Eerikäinen K, Hyyppä J, Pitkänen J (2005) Identifying and quantifying structural characteristics of heterogeneous boreal forests using laser scanner data. For Ecol Manage 216:41–50

Maltamo M, Suvanto A, Packalén P (2007b) Comparison of basal area and stem frequency diameter distribution modelling using airborne laser scanner data and calibration estimation. For Ecol Manage 247:26–34

Mamo N, Sterba H (2006) Site index functions for *Cupressus lusitanica* at Munesa Shashemene, Ethiopia. For Ecol Manage 237:429–435

Marsden C, le Maire G, Stape JL, Lo Seen D, Roupsard O, Cabral O, Epron D, Lima AMN, Nouvellon Y (2010) Relating MODIS vegetation index time-series with structure, light absorption and stem production of fast-growing *Eucalyptus* plantations. For Ecol Manage 259:1741–1753

Marshall DD, Iles K, Bell JF (2004) Using a large-angle gauge to select trees for measurement in variable plot sampling. Can J For Res 34:840–845

Martin AJ (1984) Testing volume equation accuracy with water displacement techniques. For Sci 30:41–50

Martínez-Vilalta J, Vanderklein D, Mencuccini M (2007) Tree height and age-related decline in growth in Scots pine (*Pinus sylvestris* L.). Oecologia 150:529–544

Martinis R, Socco LV, Sambuelli L, Nicolotti G, Schmitt O, Bucur V (2004) Tomographie ultrasonore pour les arbres sur pied. Ann For Sci 61:157–162

Mascaro J, Litton CM, Hughes RF, Uowolo A, Schnitzer SA (2014) Is logarithmic transformation necessary in allometry? Ten, one-hundred, *one-thousand-times* yes. Biol J Linn Soc 111:230–233

Mason NWH, Beets PN, Payton I, Burrows L, Holdaway RJ, Carswell FE (2014) Individual-based allometric equations accurately measure carbon storage and sequestration in shrublands. Forests 5:309–324

Massada AB, Carmel Y, Tzur GE, Grünzweig JM, Yakir D (2006) Assessment of temporal changes in aboveground forest tree biomass using aerial photographs and allometric equations. Can J For Res 36:2585–2594

Mate R, Johansson T, Sitoe A (2014) Biomass equations for tropical forest tree species in Mozambique. Forests 5:535–556

Mattheck C (1991) Trees, the mechanical design. Springer, Berlin

Max TA, Burkhart HE (1976) Segmented polynomial regression applied to taper equations. For Sci 22:283–289

MBAC Consulting (2003) South east Queensland private native forest inventory. Department of Agriculture, Fisheries and Forestry, Canberra

Mbow C, Verstraete MM, Sambou B, Diaw AT, Neufeldt H (2014) Allometric models for aboveground biomass in dry savanna trees of the Sudan and Sudan-Guinean ecosystems of Southern Senegal. J For Res 19:340–347

McKenney DW, Pedlar JH (2003) Spatial models of site index based on climate and soil properties for two boreal tree species in Ontario, Canada. For Ecol Manage 175:497–507

McRoberts RE, Holden GR, Nelson MD, Liknes GC, Gormanson DD (2006) Using satellite imagery as ancillary data for increasing the precision of estimates for the Forest Inventory and Analysis program of the USDA Forest Service. Can J For Res 36:2968–2980

McRoberts RE, Tomppo EO, Næsset E (2010) Advances and emerging issues in national forest inventories. Scand J For Res 25:368–381

McRoberts RE, Westfall JA (2014) Effects of uncertainty in model predictions of individual tree volume on large area volume estimates. For Sci 60:34–42

Medhurst JL, Beadle CL (2002) Sapwood hydraulic conductivity and leaf area-sapwood area relationships following thinning of a *Eucalyptus nitens* plantation. Plant Cell Environ 25:1011–1019

Medlyn BE, Berbigier P, Clement R, Grelle A, Loustau D, Linder S, Wingate L, Jarvis PG, Sigurdsson BD, McMurtrie RE (2005) Carbon balance of coniferous forests growing in contrasting climates: model-based analysis. Agric For Meteorol 131:97–124

Mehtätalo L (2006) Estimating the effect of overlapping crowns from aerial inventory estimates. Can J For Res 36:1649–1660

Mencuccini M, Martínez-Vilalta J, Hamid HA, Korakaki E, Vanderklein D (2007) Evidence for age- and size-mediated controls of tree growth form from grafting studies. Tree Physiol 27:463–473

Meng Q, Cieszewski CJ, Madden M, Borders B (2007) A linear mixed-effects model of biomass and volume of trees using Landsat ETM+ images. For Ecol Manage 244:93–101

Michelakis D, Stuart N, Lopez G, Linares V, Woodhouse IH (2014) Local-scale mapping of biomass in tropical lowland pine savannas using ALOS PALSAR. Forests 5:2377–2399

Miller TM, Allen HL, Maier CA (2006) Quantifying the coarse-root biomass of intensively managed loblolly pine plantations. Can J For Res 36:12–22

Mitchard ETA et al (2014) Markedly divergent estimates of Amazon forest carbon density from ground plots and satellites. Glob Ecol Biogeogr 23:935–946

Moisen GG, Edwards TC (1999) Use of generalized linear models and digital data in a forest inventory of northern Utah. J Agric Biol Environ Stat 4:372–390

Moisen GG, Frescino TS (2002) Comparing five modelling techniques for predicting forest characteristics. Ecol Mod 157:209–225

Mokany K, Raison RJ, Prokushkin AS (2006) Critical analysis of root: shoot ratios in terrestrial biomes. Global Change Biol 12:84–96

Monserud R, Marshall JD (1999) Allometric crown relations in three northern Idaho conifers. Can J For Res 29:521–535

Montagu KD, Düttmer K, Barton CVM, Cowie AL (2005) Developing general allometric relationships for regional estimates of carbon sequestration—an example using *Eucalyptus pilularis* from seven contrasting sites. For Ecol Manage 204:113–127

Montes F, Pita P, Rubio A, Cañellas I (2007) Leaf area index estimation in mountain even-aged *Pinus sylvestris* stands from hemispherical photographs. Agric For Meteorol 145:215–228

Montes F, Rubio A, Barbeito I, Cañellas I (2008) Characterization of the spatial structure of the canopy in *Pinus sylvestris* L. stands in Central Spain from hemispherical photographs. For Ecol Manage 255:580–590

Mooney SJ, Pridmore TP, Helliwell J, Bennett MJ (2012) Developing X-ray Computed Tomography to non-invasively image 3-D root systems architecture in soil. Plant Soil 352:1–22

Moradi AB, Conesa HM, Robinson B, Lehmann E, Kuehne G, Kaestner A, Oswald S, Schulin R (2009) Neutron radiography as a tool for revealing root development in soil: capabilities and limitations. Plant Soil 318:243–255

Morgenroth J, Gomez C (2014) Assessment of tree structure using a 3D image analysis technique—a proof of concept. Urban For Urban Green 13:198–203

Morley T, Little K (2012) Comparison of taper functions between two planted and coppiced eucalypt clonal hybrids, South Africa. New For 43:129–141

Morton RT, Titus SJ, Bonnor GM, Grabowski TI (1990) An assessment of white spruce tree volume equations in Canada. For Chron 66:600–605

Moser JW (1972) Dynamics of an uneven-aged forest stand. For Sci 18:184–191

Mosseler A, Major JE, Labrecque M, Larocque GR (2014) Allometric relationships in coppice biomass production for two North American willows (*Salix* spp.) across three different sites. For Ecol Manage 320:190–196

Muhairwe CK (2000) Bark thickness equations for five commercial tree species in regrowth forests of Northern New South Wales. Aust For 63:34–43

Muinonen E, Pitkänen J, Hung NP, Tinh MV, Eerikäinen K (2014) Integrating multi-source data for a tropical forest inventory—a case study in the Kon Tum region, Vietnam. Aust For 77:92–104

Mummery D, Battaglia M (2001) Applying PROMOD spatially across Tasmania with sensitivity analysis to screen for prospective *Eucalyptus globulus* plantation sites. For Ecol Manage 140:51–63

Muukkonen P (2007) Generalized allometric volume and biomass equations for some tree species in Europe. Eur J For Res 126:157–166

Myers BJ, Benson ML (1981) Rainforest species on large-scale color photos. Photogram Eng Remote Sens 47:505–513

Myers BJ, Benson ML, Craig IE, Wear JF, West PW (1984) Shadowless or sunlit photos for forest disease detection. Photogram Eng Remote Sens 50:53–72

Nabeshima E, Hiura T (2008) Size-dependency in hydraulic and photosynthetic properties of three *Acer* species having different maximum sizes. Ecol Res 23:281–288

Næsset E (2004) Accuracy of forest inventory using airborne laser scanning: evaluating the first Nordic full-scale operational project. Scand J For Res 19:554–557

Næsset E (2007) Airborne laser scanning as a method in operational forest inventory: status of accuracy assessments accomplished in Scandinavia. Scand J For Res 22:433–442

Nakajima H, Kume A, Ishida M, Ohmiya T, Mizoue N (2011) Evaluation of estimates of crown condition in forest monitoring: comparison between visual estimation and automated crown image analysis. Ann For Sci 68:1333–1340

National Forest Inventory (1998) Australia's sate of the forest report 1998. Bureau of Rural Sciences, Commonwealth of Australia, Canberra

Návar J (2009) Allometric equations for tree species and carbon stocks for forests of northwestern Mexico. For Ecol Manage 257:427–434

Návar J, de Jesús Rodríguez-Flores F, Domínguez-Calleros PA (2013) Taper functions and merchantable timber for temperate forests of northern Mexico. Ann For Res 56:165–178

Newton PF, Sharma M (2008) Evaluation of sampling design on taper equation performance in plantation-grown *Pinus banksiana*. Scand J For Res 23:358–370

Ngomanda A et al (2014) Site-specific versus pantropical allometric equations: which option to estimate the biomass of a moist central African forest? For Ecol Manage 312:1–9

Ngomanda A, Mavouroulou QM, Obiang NLE, Iponga DM, Mavoungou JF, Lepengue N, Picard N, Mbatchi B (2012) Derivation of diameter measurements for buttressed trees, an example from Gabon. J Trop Ecol 28:299–302

Nieschulze J, Erasmi S, Dietz J, Hölscher D (2009) Satellite-based prediction of rainfall interception by tropical forest stands of a human-dominated landscape in Central Sulawesi, Indonesia. J Hydrol 364:227–235

Nigh G, Smith W (2012) Effect of climate on lodgepole pine stem taper in British Columbia, Canada. Forestry 85:579–587

Nightingale JM, Hill MJ, Phinn SR, Davies ID, Held AA (2008a) Use of 3-PG and 3-PGS to simulate forest growth dynamics of Australian tropical rainforests. II. An integrated system for modelling forest growth and scenario assessment within the wet tropics bioregion. For Ecol Manage 254:122–133

Nightingale JM, Hill MJ, Phinn SR, Davies ID, Held AA, Erskine PD (2008b) Use of 3-PG and 3-PGS to simulate forest growth dynamics of Australian tropical rainforests. I. Parameterisation and calibration for old-growth, regenerating and plantation forests. For Ecol Manage 254:107–121

Niklas KJ (2005) Modelling below- and above-ground biomass for non-woody and woody plants. Ann Bot 95:315–321

Nogueira EM, Fearnside PM, Nelson BW, Barbosa RI, Keizer EWH (2008) Estimates of forest biomass in the Brazilian Amazon: new allometric equations and adjustments to biomass from wood-volume inventories. For Ecol Manage 256:1853–1867

Nolé A, Law BE, Magnani F, Matteucci G, Ferrara A, Ripullone F, Borghetti M (2009) Application of the 3-PGS model to assess carbon accumulation in forest ecosystems at a regional level. Can J For Res 39:1647–1661

Nord-Larsen T (2006) Developing dynamic site index curves for European beech (*Fagus sylvatica* L.) in Denmark. For Sci 52:173–181

O'Grady AP, Worledge D, Battaglia M (2006) Above- and below-ground relationships, with particular reference to fine roots, in a young *Eucalyptus globulus* (Labill.) stand in southern Tasmania. Trees 20:531–538

Ometto JP, Aguiar AP, Assis T, Soler L, Valle P, Tejada G, Lapola DM, Meir P (2014) Amazon forest biomass density maps: tackling the uncertainty in carbon emission estimates. Clim Change 124:545–560

Opie JE (1976) Volume functions for trees of all sizes. Forestry Commission of Victoria, Forestry Technical Papers 25:27–30

Osler GHR, West PW, Downes GM (1996a) Effects of bending stress on taper and growth of stems of young *Eucalyptus regnans* trees. Trees 10:239–246

Osler GHR, West PW, Laffan MD (1996b) Test of a system to predict productivity of eucalypt plantations in Tasmania. Aust For 59:57–63

Özçelik R (2008) Comparison of formulae for estimating tree bole volumes of *Pinus sylvestris*. Scand J For Res 23:412–418

Özçelik R, Diamantopoulou MJ, Brooks JR (2014) The use of tree crown variables in over-bark diameter and volume prediction models. iForest 7:132–139

Özçelik R, Diamantopoulou MJ, Brooks JR, Wiant HV (2010) Estimating tree bole volume using artificial neural network models for four species in Turkey. J Environ Manage 91:742–753

Özçelik R, Diamantopoulou MJ, Wiant HV, Brooks JR (2008a) Comparative study of standard and modern methods for estimating tree bole volume of three species in Turkey. For Prod J 58:73–81

Özçelik R, Wiant HV, Brooks JR (2008b) Accuracy using xylometry of log volume estimates for two tree species in Turkey. Scand J For Res 23:272–277

Pajtík J, Konôpka B, Lukac M (2008) Biomass functions and expansion factors in young Norway spruce (*Picea abies* [L.] Karst) trees. For Ecol Manage 256:1096–1103

Pandey CB, Singh L, Singh SK (2011) Buttresses induced habitat heterogeneity increases nitrogen availability. For Ecol Manage 262:1679–1685

Paneque-Gálvez J, McCall MK, Napoletano BM, Wich SA, Koh LP (2014) Small drones for community-based forest monitoring: an assessment of their feasibility and potential in tropical areas. Forests 5:1481–1507

Parresol BR (1999) Assessing tree and stand biomass: a review with examples and critical comparisons. For Sci 45:573–593

Parveaud C-E, Chopard J, Dauzat J, Courbaud B, Auclair D (2008) Modelling foliage character-istics in 3D tree crowns: influence on light interception and leaf irradiance. Trees 22:87–104

Pascual C, García-Abril A, García-Mortem LG, Martín-Fernández S, Cohen WB (2008) Object-based semi-automatic approach for forest structure characterization using lidar data in hetero-geneous *Pinus sylvestris* stands. For Ecol Manage 255:3677–3685

Patankar R, Thomas SC, Smith SM (2011) A gall-inducing arthropod drives declines in canopy tree photosynthesis. Oecologia 167:701–709

Paul KI, Jacobsen K, Koul V, Leppert P, Smith J (2008) Predicting growth and sequestration of carbon by plantations growing in regions of low-rainfall in southern Australia. For Ecol Manage 254:205–216

Payendah B, Ek AR (1986) Distance methods and density estimators. Can J For Res 16:918–924 (see also the erratum in Vol. 17:95–96, 1987)

Peichl M, Arain MA (2007) Allometry and partitioning of biomass in an age-sequence of white pine forests. For Ecol Manage 253:68–80

Peng C (2000a) Growth and yield models for uneven-aged stands: past, present and future. For Ecol Manage 132:259–279

Peng C (2000b) Understanding the role of forest simulation models in sustainable forest manage-ment. Environ Impact Assess Rev 20:481–501

Pereira JMC, Tomé M, Carreiras JMB, Tomé JA, Pereira JS, David JS, Fabião MD (1997) Leaf area estimation from tree allometrics in *Eucalyptus globulus* plantations. Can J For Res 27:166–173

Pérez Cordero LD, Kanninen M (2003) Aboveground biomass of *Tectona grandis* plantations in Costa Rica. J Trop For Sci 15:199–213

Pérez-Harguindeguy N et al (2013) New handbook for standardised measurement of plant functional traits worldwide. Aust J Bot 61:167–234

Persson Å, Holmgren J, Söderman U (2002) Detecting and measuring individual trees using an airborne laser scanner. Photogram Eng Remote Sens 68:925–932

Pesonen A, Maltamo M, Eerikäinen K, Packalén P (2008) Airborne laser scanning-based predic-tion of coarse woody debris volumes in a conservation area. For Ecol Manage 255:3288–3296

Peuhkurinen J, Maltamo M, Malinen J, Pitkänen J, Packalén P (2007) Preharvest measurement of marked stands using airborne laser scanning. For Sci 53:653–661

Philip MS (1994) Measuring trees and forests, 2nd edn. CAB International, Oxford

Picard N, Henry M, Mortier F, Trotta C, Saint-André L (2012) Using Bayesian model averaging to predict tree aboveground biomass in tropical moist forests. For Sci 58:15–23

Picard N, Kouyaté AM, Dessard H (2005) Tree density estimation using a distance method in Mali savanna. For Sci 51:7–18

Pierzchala M, Talbot B, Astrup R (2014) Estimating soil displacement from timber extraction trails in steep terrain: application of an unmanned aircraft for 3D modeling. Forests 5:1212–1223

Pilli R, Anfodillo T, Carrer M (2006) Towards a functional and simplified allometry for estimating forest biomass. For Ecol Manage 237:583–593

Piper FI, Fajardo A (2011) No evidence of carbon limitation with tree age and height in *Nothofagus pumilio* under Mediterranean and temperate climate conditions. Ann Bot 108:907–917

Pitt DG, Bell FW (2004) Effects of stand tending on the estimation of aboveground biomass of planted juvenile white spruce. Can J For Res 34:649–658

Planck NRV, MacFarlane DW (2014) Modelling vertical allocation of tree stem and branch volume for hardwoods. Forestry 87:459–469

Pocewicz AL, Gessler P, Robinson AP (2004) The relationship between effective plant area index and Landsat spectral response across elevation, solar insolation, and spatial scales in a northern Idaho forest. Can J For Res 34:465–480

Pohjonen VM (1991) Volume equations and volume tables of *Juniperus procera* Hocht. ex. Endl. For Ecol Manage 44:185–200

Pokharel B, Froese RE (2009) Representing site productivity in the basal area increment model for FVS-Ontario. For Ecol Manage 258:657–666

Poorter L, Rozendaal DMA (2008) Leaf size and leaf display of thirty-eight tropical tree species. Oecologia 158:35–46

Popescu SC (2007) Estimating biomass of individual pine trees using airborne lidar. Biomass Bioenergy 31:646–655

Popescu SC, Wynne RH (2004) Seeing the trees in the forest: using Lidar and multispectral data fusion with local filtering and variable window size for estimating tree height. Photogram Eng Remote Sens 70:589–604

Popescu SC, Wynne RH, Scrivani JA (2004) Fusion of small-footprint lidar and multispectral data to estimate plot-level volume and biomass in deciduous and pine forests in Virginia, USA. For Sci 50:551–565

Porté A, Bartelink HH (2002) Modelling mixed forest growth: a review of models for forest management. Ecol Mod 150:141–188

Pothier D, Margolis HA, Waring RH (1989) Patterns of change of saturated sapwood permeability and sapwood conductance with stand development. Can J For Res 19:432–439

Pretzsch H, Biber P (2005) A re-evaluation of Reineke's rule and stand density index. For Sci 51:304–320

Pretzsch H, Biber P, Ďurský J, von Gadow K, Hasenauer H, Kändler G, Kenk G, Kublin E, Nagei E, Pukkala T, Skovsgaard JP, Sodtke R, Sterba H (2002) Recommendations for standardized documentation and further development of forest growth simulators. Forstw Cbl 121:138–151

Pretzsch H, Dauber E, Biber P (2013) Species-specific and ontogeny-related stem allometry of European forest trees: evidence from extensive stem analyses. For Sci 59:290–302

Pulkkinen M (2012) On non-circularity of tree stem cross-sections: effect of diameter selection on cross-section area estimation, Bitterlich sampling and stem volume estimation in Scots pine. Silva Fenn 46:747–986

Räim O, Kaurilind E, Hallik L, Merilo E (2012) Why does needle photosynthesis decline with tree height in Norway spruce? Plant Biol 14:306–314

Rhoads AG, Hamburg SP, Fahey TJ, Siccama TG, Kobe R (2004) Comparing direct and indirect methods of assessing canopy structure in a northern hardwood forest. Can J For Res 34:584–591

Riaño D, Valladares F, Condés S, Chuvieco E (2004) Estimation of leaf area index and covered ground from airborne laser scanner (Lidar) in two contrasting forests. Agric For Meteorol 124:269–275

Richards GP, Brack C (2004) A continental biomass and stock change estimation approach for Australia. Aust For 67:284–288

Richardson B, Watt MS, Mason EG, Kriticos DJ (2006) Advances in modelling and decision support systems for vegetation management in young forest plantations. Forestry 79:29–42

Roberts SD, Dean TJ, Evans DL, McCombs JW, Harrington RL, Glass PA (2005) Estimating individual tree leaf area in loblolly pine plantations using LiDAR-derived measurements of height and crown dimensions. For Ecol Manage 213:54–70

Robertson A (1991) Centroid of wood density, bole eccentricity, and tree-ring width in relation to vector winds in wave forests. Can J For Res 21:73–82

Robinson D (2004) Scaling the depths: below-ground allocation in plants, forests and biomes. Funct Ecol 18:290–295

Rock J (2007) Suitability of published biomass equations for aspen in Central Europe—results from a case study. Biomass Bioenergy 31:299–307

Rodríguez F, Lizarralde I, Bravo F (2013) Additivity on nonlinear stem taper functions: a case for Corsican pine in northern Spain. For Sci 59:464–471

Roxburgh SH, Barrett DJ, Berry SL, Carter JO, Davies ID, Gifford RM, Kirschbaum MUF, McBeth BP, Noble IR, Parton WG, Raupach MR, Roderick ML (2004) A critical overview of model estimates of net primary productivity for the Australian continent. Funct Plant Biol 31:1043–1059

Running SW (1994) Testing FOREST-BGC ecosystem process simulations across a climatic gradient in Oregon. Ecol Appl 4:238–247

Rustagi KP, Loveless RS (1990) Improved cubic volume prediction using a new measure of form factor. For Ecol Manage 40:1–11

Ryan M (2013) Adaptive silviculture in regrowth eucalypt forests in Victoria and the implications for water, wood, wildlife and wildfire. Aust For 76:173–182

Ryan MG, Phillips N, Bond BJ (2006) The hydraulic limitation hypothesis revisited. Plant Cell Environ 29:367–381

Ryan PJ, Harper RJ, Laffan M, Booth TH, McKenzie NJ (2002) Site assessment for farm forestry in Australia and its relationship to scale, productivity and sustainability. For Ecol Manage 171:133–152

Rytter L (2006) A management regime for hybrid aspen stands combining conventional forestry techniques with early biomass harvests to exploit their early rapid growth. For Ecol Manage 236:422–426

Rytter R-M (1999) Fine-root production and turnover in a willow plantation estimated by different calculation methods. Scand J For Res 14:526–537

Ryu Y, Sonnentag O, Nilson T, Vargas R, Kobayashi H, Wenk R, Baldocchi DD (2010) How to quantify tree leaf area index in an open savanna ecosystem: a multi-instrument and multi-model approach. Agric For Meteorol 150:63–76

Sabatia CO, Burkhart HE (2014) Predicting site index of plantation loblolly pine from biophysical variables. For Ecol Manage 326:142–156

Saenger P (2002) Mangrove ecology, silviculture and conservation. Kluwer, Dordrecht

Saint-André L, M'Bou AM, Mabiala A, Mouvondy W, Jourdan C, Roupsard O, Deleporte P, Hamel O, Nouvellon Y (2005) Age-related equations for above- and below-ground biomass of a *Eucalyptus* hybrid in Congo. For Ecol Manage 205:199–214

Šálek L, Zahradník D, Marušák R, Jeřábková L, Merganič J (2013) Forest edges in managed riparian forests in the eastern part of the Czech Republic. For Ecol Manage 305:1–10

Sanchez-Azofeifa GA, Castro-Esau KL, Kurz WA, Joyce A (2009) Monitoring carbon stocks in the tropics and the remote sensing operational limitations: from local to regional projects. Ecol Appl 19:480–494

Sands PJ, Battaglia M, Mummery D (2000) Application of process-based models to forest management: experience with PROMOD, a simple plantation productivity model. Tree Physiol 20:383–392

Sandström F, Petersson H, Kruys N, Ståhl G (2007) Biomass conversion factors (density and carbon concentration) by decay classes for dead wood of *Pinus sylvestris*, *Picea abies* and *Betula* spp. in boreal forests of Sweden. For Ecol Manage 243:19–27

Saremi H, Kumar L, Turner R, Stone C (2014a) Airborne LiDAR derived canopy height model reveals a significant difference in radiata pine (*Pinus radiata* D. Don) heights based on slope and aspect of sites. Trees 28:733–744

Saremi H, Kumar L, Turner R, Stone C, Melville G (2014b) Impact of local slope and aspect assessed from LiDAR records on tree diameter in radiata pine (*Pinus radiata* D. Don) plantations. Ann For Sci 71:771–780

Satterthwaite FE (1946) An approximate distribution of estimates of variance components. Biometrics Bull 2:110–114

Schleppi P, Conedera M, Sedivy I, Thimonier A (2007) Correcting non-linearity and slope effects in the estimation of the leaf area index of forests from hemispherical photographs. Agric For Meteorol 144:236–242

Schmid-Haas P (1969) Stichproben am waldrand. Mitt Schweiz Anst Forstl Versuchswes 45:234–303

Schmitt DC, Grigal DF (1981) Generalized biomass estimation equations for *Betula papyrifera* Marsh. Can J For Res 11:837–840

Schneider R, Riopel M, Pothier D, Cote L (2008) Predicting decay and round-wood end use volume in trembling aspen (*Populus tremuloides* Michx.). Ann For Sci 65:608–622

Schreuder HT, Gregoire TG, Wood GB (1993) Sampling methods for multiresource forest inventory. Wiley, New York

Schreyer J, Tigges J, Lakes T, Churkina G (2014) Using airborne LiDAR and Quickbird data for modelling urban tree carbon storage and its distribution—a case study of Berlin. Remote Sens 6:10636–10655

Seidel D, Fleck S, Leuschner C (2012) Analyzing forest canopies with ground-based laser scanning: a comparison with hemispherical photography. Agric For Meteorol 154:1–8

Sharma M, Burkhart HE (2003) Selecting a level of conditioning for the segmented polynomial taper equation. For Sci 49:324–330

Sharma M, Oderwald RG (2001) Dimensionally compatible volume and taper equations. Can J For Res 31:797–803

Sharma M, Parton J (2009) Modeling stand density effects on taper for Jack pine and black spruce plantations using dimensional analysis. For Sci 55:268–282

Shelburne VB, Hedden RL, Allen RM (1993) The effects of site, stand density, and sapwood permeability on the relationship between leaf area and sapwood area in loblolly pine (*Pinus taeda* L.). For Ecol Manage 58:193–209

Shiver BD, Borders BE (1996) Sampling techniques for forest resource inventory. Wiley, New York

Shiver BD, Brister GH (1992) Tree and stand volume functions for *Eucalyptus saligna*. For Ecol Manage 47:211–223

Sims REH, Senelwa K, Maiava T, Bullock BT (1999) *Eucalyptus* species for biomass energy in New Zealand—part II: coppice performance. Biomass Bioenergy 17:333–343

Skovsgaard JP, Vanclay JK (2008) Forest site productivity: a review of the evolution of dendrometric concepts for even-age stands. Forestry 81:13–31

Skovsgaard JP, Vanclay JK (2013) Forest site productivity: a review of spatial and temporal variability in natural site conditions. Forestry 86:305–315

Smith B, Knorr W, Widlowski J-L, Pinty B, Gobron N (2008) Combining remote sensing data with process modelling to monitor boreal conifer forest carbon balances. For Ecol Manage 255:3985–3994

Snowdon P (2001) Short-term predictions of growth of *Pinus radiata* with models incorporating indices of annual climatic variation. For Ecol Manage 152:1–11

Snowdon P, Eamus D, Gibbons P, Khanna PK, Keith H, Raison RJ, Kirschbaum MUF (2000) Synthesis of allometrics, review of root biomass and design of future woody biomass sampling strategies. National Carbon Accounting System Technical Report No. 17. Australian Greenhouse Office, Canberra

Snowdon P, Raison J, Keith H, Ritson P, Grierson P, Adams M, Montagu K, Bi H, Burrows W, Eamus D (2002) Protocol for sampling tree and stand biomass. Technical Report No. 31. Australian Greenhouse Office, Canberra

Soares P, Tomé M (2002) Height-diameter equation for first rotation eucalypt plantations in Portugal. For Ecol Manage 166:99–109

Sochacki SJ, Harper RJ, Smetten KRJ (2007) Estimation of woody biomass production from a short-rotation bio-energy system in semi-arid Australia. Biomass Bioenergy 31:608–616

Specht A, West PW (2003) Estimation of biomass and sequestered carbon on farm forest plantations in northern New South Wales, Australia. Biomass Bioenergy 25:363–379

Specht RL, Specht A (1999) Australian plant communities. Oxford University Press, Melbourne

Sperry JS, Meinzer FC, McCulloh KA (2008) Safety and efficiency conflicts in hydraulic architecture: scaling from tissues to trees. Plant Cell Environ 31:632–645

Ståhl G, Holm S, Gregoire TG, Gobakken T, Næsset E, Nelson R (2011) Model-abased inference for biomass estimation in a LIDAR sample survey in Hedmark County, Norway. Can J For Res 41:96–107

Stancioiu PT, O'Hara KL (2005) Sapwood-area—leaf area relationships for coast redwood. Can J For Res 35:1250–1255

Stephens PR, Kimberley MO, Beets PN, Paul TSH, Searles N, Bell A, Brack C, Broadley J (2012) Airborne scanning LiDAR in a double sampling forest carbon inventory. Remote Sens Environ 117:348–357

Stone C, Carnegie A, Melville G, Smith D, Nagel M (2013) Aerial mapping canopy damage by the aphid *Essigella californica* in a *Pinus radiata* plantation in southern New South Wales: what are the challenges? Aust For 76:101–109

St-Onge B, Jumulet J, Cobello M, Véga C (2004) Measuring individual tree height using a combination of stereophotogrammetry and lidar. Can J For Res 34:2122–2130

St-Onge B, Treitz P, Wulder MA (2003) Tree canopy height estimation with scanning lidar. In: Wulder MA, Franklin SE (eds) Remote sensing of forest environments: concepts and case studies. Kluwer, Dordrecht, pp 489–509

Straub C, Tian JJ, Seitz R, Reinartz P (2013) Assessment of Cartosat-1 and WorldView-2 stereo imagery in combination with a LiDAR-DTM for timber volume estimation in a highly structured forest in Germany. Forestry 86:463–473

Suganuma H, Abe Y, Taniguchi M, Tanouchi H, Utsugi H, Kojima T, Yamada K (2006) Stand biomass estimation method by canopy coverage for application to remote sensing in an arid area of Western Australia. For Ecol Manage 222:75–87

Swenson JJ, Waring RH, Fan W, Coops N (2005) Predicting site index with a physiologically based growth model across Oregon, USA. Can J For Res 35:1697–1707

Tanaka T, Park H, Hattori S (2003) Distinguishing foliage from branches in the non-destructive measurement of the three-dimensional structure of mountain forest canopies. For Chron 79:313–317

Tanaka T, Park H, Hattori S (2004) Measurement of forest canopy structure by a laser plane rangfinding method. Improvement of radiative resolution and examples of its application. Agric For Meteorol 125:129–142

Tarp-Johansen MJ (2002a) Automatic stem mapping in three dimensions by template matching from aerial photographs. Scand J For Res 17:359–368

Tarp-Johansen MJ (2002b) Stem diameter estimation from aerial photographs. Scand J For Res 17:369–376

Tatarinov F, Urban J, Čermák J (2008) Application of "clump technique" for root system studies of *Quercus robur* and *Fraxinus excelsior*. For Ecol Manage 255:495–505

Tatarinov FA, Cienciala E (2006) Application of BIOME-BGC model to managed forests. 1. Sensitivity analysis. For Ecol Manage 237:267–279

Temesgen H, Monleon V, Weiskittel A, Wilson D (2011) Sampling strategies for efficient estimation of tree foliage biomass. For Sci 57:153–163

Ter-Mikaelian MT, Korzukhin MD (1997) Biomass equations for sixty-five North American tree species. For Ecol Manage 97:1–24

Ter-Mikaelian MT, Zakrzewski WT, MacDonald GB, Weingartner DH (2004) Stem profile equations for young trembling aspen in northern Ontario. Ann For Sci 61:109–115

Teshome T (2005) A ratio method for predicting stem merchantable volume and associated taper equations for *Cupressus lusitanica*, Ethiopia. For Ecol Manage 204:171–179

Thomas SC, Martin AR (2012) Carbon content of tree tissues: a synthesis. Forests 3:332–352

Tickle PK, Lee A, Lucas RM, Austin J, Witte C (2006) Quantifying Australian forest floristics and structure using small footprint LiDAR and large scale aerial photography. For Ecol Manage 223:379–394

Tobin B, Black K, Osborne B, Reidy B, Bolger T, Nieuwenhuis M (2006) Assessment of allometric algorithms for estimating leaf biomass, leaf area index and litter fall in different-aged Sitka spruce forests. Forestry 79:453–465

Tomppo E, Malimbwi R, Katila M, Mäkisara K, Henttonen HM, Chamuya N, Zahabu E, Otieno J (2014) A sampling design for a large area forest inventory: case Tanzania. Can J For Res 44:931–948

Turvey ND, Booth TH, Ryan PJ (1990) A soil technical classification system for *Pinus radiata* (D. Don) plantations. II. A basis for estimation of crop yield. Aust J Soil Res 28:813–824

Ung C-H, Bernier P, Guo X-J (2008) Canadian national biomass equations: new parameter estimates that include British Columbia data. Can J For Res 38:1123–1132

Ung CH, Guo XJ, Fortin M (2013) Canadian national taper models. For Chron 89:211–224

Uzoh FCC (2001) A height increment equation for young ponderosa pine plantations using precipitation and soil factors. For Ecol Manage 142:193–203

Vadeboncoeur MA, Hamburg SP, Yanai RD (2007) Validation and refinement of allometric equations for roots of northern hardwoods. Can J For Res 37:1777–1783

Valentine HT, Affleck DLR, Gregoire TG (2009) Systematic sampling of discrete and continuous populations: sample selection and the choice of estimator. Can J For Res 39:1061–1068

Valentine HT, Gregoire TG (2001) A switching model of bole taper. Can J For Res 31:1400–1409

Valentine HT, Mäkelä A (2005) Bridging process-based and empirical approaches to modeling tree growth. Tree Physiol 25:769–779

Valentine HT, Tritton LM, Furnival GM (1984) Subsampling trees for biomass, volume, or mineral content. For Sci 30:673–681

Valinger E (1992) Effects of wind sway on stem form and crown development of Scots pine (*Pinus sylvestris* L.). Aust For 55:15–21

Vallet P, Dhôte J-F, Le Moguédec G, Ravart M, Pignard G (2006) Development of total aboveground volume equations for seven important forest tree species in France. For Ecol Manage 229:98–110

van Breugel M, Ransijn J, Craven D, Bongers F, Hall JS (2011) Estimating carbon stock in secondary forests: decisions and uncertainties associated with allometric biomass models. For Ecol Manage 262:1648–1657

Van Camp N, Vande Walle I, Mertens J, De Neve S, Samson R, Lust N, Lemeur R, Boeckx P, Lootens P, Beheydt D, Mestdagh I, Sleutel S, Verbeeck H, Van Cleemput O, Hofman G, Carlier L (2004) Inventory-based carbon stock of Flemish forests: a comparison of European biomass expansion factors. Ann For Sci 61:677–682

Van Deusen PC (1987) Combining taper functions and critical height sampling for unbiased stand volume estimation. Can J For Res 17:1416–1420

Van Deusen PC (1994) Alternative volume and quantity formulas for individual trees. Can J For Res 24:50–52

Van Deusen PC, Lynch TB (1987) Efficient unbiased tree-volume estimation. For Sci 33:583–590

van Laar A, Akça A (2007) Forest mensuration. Springer, Dordrecht

van Leeuwen M, Hilker T, Coops NC, Frazer G, Wulder MA, Newnham GJ, Culvenor DS (2011) Assessment of standing wood and fiber quality using ground and airborne laser scanning: a review. For Ecol Manage 261:1467–1478

van Mantgem PJ, Stephenson NL (2004) Does coring contribute to tree mortality? Can J For Res 34:2394–2398

Van Tuyl S, Law BE, Turner DP, Gitelman AI (2005) Variability in net primary production and carbon storage in biomass across Oregon forests—an assessment integrating data from forest inventories, intensive sites, and remote sensing. For Ecol Manage 209:273–291

Vanclay J (1992) Assessing site productivity in tropical moist forest: a review. For Ecol Manage 54:257–287

Vanclay JK (1995) Growth models for tropical forests: a synthesis of models and methods. For Sci 41:7–42

Vanclay JK (2009) Managing water use from forest plantations. For Ecol Manage 257:385–389

Vanderklein D, Martínez-Vilalta J, Lee S, Mencuccini M (2007) Plant size, not age, regulates growth and gas exchange in grafted Scots pine trees. Tree Physiol 27:71–79

VanderSchaaf CL, Burkhart HE (2007) Comparison of methods to estimate Reineke's maximum size-density relationship species boundary line slope. For Sci 53:435–442

Vanninen P, Mäkelä A (1999) Fine root biomass of Scots pine stands differing in age and soil fertility in southern Finland. Tree Physiol 19:823–830

Verwijst T, Telenius B (1999) Biomass estimation procedures in short rotation forestry. For Ecol Manage 121:137–146

Vieilledent G, Vaudry R, Andriamanohisoa SFD, Rakotonarivo OS, Randrianasolo HZ, Razafindrabe HN, Bidaud Rakotoarivony C, Ebeling J, Rasamoelina M (2012) A universal approach to estimate biomass and carbon stock in tropical forests using generic allometric models. Ecol Appl 22:572–583

Vincent G, Sabatier D, Rutishauser E (2014) Revisiting a universal airborne light detection and ranging approach for tropical forest carbon mapping: scaling-up from tree to stand to landscape. Oecologia 175:439–443

Vogt KA, Vogt DJ, Bloomfield J (1998) Analysis of some direct and indirect methods for estimating root biomass and production of forests at an ecosystem level. Plant Soil 200:71–89

Volcani A, Karnieli A, Svoray T (2005) The use of remote sensing and GIS for spatio-temporal analysis of the physiological state of a semi-arid forest with respect to drought years. For Ecol Manage 215:239–250

Wang C (2006) Biomass allometric equations for 10 co-occurring tree species in Chinese temperate forests. For Ecol Manage 222:9–16

Wang C, Lu Z, Haithcoat TL (2007a) Using Landsat images to detect oak decline in the Mark Twain National Forest, Ozark Highlands. For Ecol Manage 240:70–78

Wang J-S, Grimley DA, Xu C, Dawson JO (2008a) Soil magnetic susceptibility reflects soil moisture regimes and the adaptability of tree species to these regimes. For Ecol Manage 255:1664–1673

Wang Q, Preda M, Cox M, Bubb K (2007b) Spatial model of site index based on γ-ray spectrometry and a digital elevation model for two *Pinus* species in Tuan Toolara State Forest, Queensland, Australia. Can J For Res 37:2299–2312

Wang X, Fang J, Zhu B (2008b) Forest biomass and root-shoot allocation in northeast China. For Ecol Manage 255:4007–4020

Waring RH, Coops NC, Landsberg JJ (2010) Improving predictions of forest growth using the 3-PGS model with observations made by remote sensing. For Ecol Manage 259:1722–1729

Waring RH, Coops NC, Mathys A, Hilker T, Latta G (2014) Process-based modeling to assess the effects of recent climatic variation on site productivity and forest function across western North America. Forests 5:518–534

Waring RH, Coops NC, Ohmann JL, Sarr DA (2002) Interpreting woody plant richness from seasonal ratios of photosynthesis. Ecology 83:2964–2970

Watt MS, Adams T, Marshall H, Pont D, Lee J, Crawley D, Watt P (2013a) Modelling variation in Pinus radiata stem volume and outerwood stress-wave velocity from LiDAR metrics. N Z J For Sci 43:1

Watt MS, Meredith A, Watt P, Gunn A (2013b) Use of LiDAR to estimate stand characteristics for thinning operations in young Douglas-fir plantations. N Z J For Sci 43:18

Watt MS, Moore JR, Façon J-P, Downes GM, Clinton PW, Coker G, Davis MR, Simcock R, Parfitt RL, Dando J, Mason EG, Bown HE (2006a) Modelling the influence of stand structural, edaphic and climatic influences on juvenile *Pinus radiata* dynamic models of elasticity. For Ecol Manage 229:136–144

Watt MS, Moore JR, Façon J-P, Downes GM, Clinton PW, Coker G, Davis MR, Simcock R, Parfitt RL, Dando J, Mason EG, Bown HE (2006b) Modelling environmental variation in Young's modulus for *Pinus radiata* and implications for determination of critical buckling height. Ann Bot 98:765–775

Weiskittel AR, Hann DW, Kershaw JA, Vanclay JK (2011) Forest growth and yield modeling. Wiley-Blackwell, Oxford

Weiss M, Baret F, Smith GJ, Jonckheere I, Coppin P (2004) Review of methods for in situ leaf area index (LAI) determination. Part II. Estimation of LAI, errors and sampling. Agric For Meteorol 121:37–53

West PW (1979) Estimation of height, bark thickness and plot volume in regrowth eucalypt forest. Aust For Res 9:295–308

West PW (1983) Comparison of stand density measures in even-aged regrowth eucalypt forest of southern Tasmania. Can J For Res 13:22–31

West PW (2005) An alternative approach to selecting a 3P sample. Paper to a meeting of the Western Forest Mensurationists, Hilo, Hawaii, July 2005.

West PW (2011) Potential for wider application of 3P sampling in forest inventory. Can J For Res 41:1500–1508

West PW (2014a) Growing plantation forests, 2nd edn. Springer, Switzerland

West PW (2014b) Precision of inventory using different edge overlap methods. Can J For Res 43:1081–1083

West PW, Beadle CL, Turnbull CRA (1989) Mechanistically based, allometric models to predict tree diameter and height in even-aged monoculture of *Eucalyptus regnans* F. Muell. Can J For Res 19:270–273

West PW, Cawsey EM, Stol J, Freudenberger D (2008) Firewood harvest from forests of the Murray-Darling Basin, Australia. Part 1. Long-term, sustainable supply available from native forests. Biomass Bioenergy 32:1206–1219

West PW, Mattay JP (1993) Yield prediction models and comparative growth rates for six eucalypt species. Aust For 56:211–225

West PW, Wells KF (1990) Estimation of leaf weight of standing trees of *Eucalyptus regnans*. Can J For Res 20:1732–1738

West PW, Wells KF, Cameron DM, Rance SJ, Turnbull CRA, Beadle CL (1991) Predicting tree diameter and height from above-ground biomass for four eucalypt species. Trees 5:30–35

Westfall JA, Scott CT (2010) Taper models for commercial tree species in the northeastern United States. For Sci 56:515–528

White JC, Wulder MA, Vastaranta M, Coops NC, Pitt D, Woods M (2013) The utility of image-based point clouds for forest inventory: a comparison with airborne laser scanning. Forests 4:518–536

Whitehead D, Edwards WRN, Jarvis PG (1984) Conducting sapwood area, foliage area, and permeability in mature trees of *Picea sitchensis* and *Pinus contorta*. Can J For Res 14:940–947

Wiant HV, Spangler ML, Baumgras JE (2002) Comparison of estimates of hardwood bole volume using importance sampling, the centroid method, and some taper equations. North J Appl For 19:141–142

Wiant HV, Wood GB, Forslund RR (1991) Comparison of centroid and paracone estimates of tree volume. Can J For Res 21:714–717

Wiant HV, Wood GB, Furnival GM (1992a) Estimating log volume using the centroid position. For Sci 38:187–191

Wiant HV, Wood GB, Gregoire TG (1992b) Practical guide for estimating the volume of a standing sample tree using either importance or centroid sampling. For Ecol Manage 49:333–339

Wiant HV, Wood GB, Miles JA (1989) Estimating the volume of a radiata pine stand using importance sampling. Aust For 52:286–292

Wiant HV, Wood GB, Williams M (1996) Comparison of three modern methods for estimating volume of sample trees using one or two diameter measurements. For Ecol Manage 83:13–16

Wiemann MC, Williamson GB (2012) Testing a novel method to approximate wood specific gravity of trees. For Sci 58:577–591

Williams MS, Gregoire TG (1993) Estimating weights when fitting linear regression models for tree volume. Can J For Res 23:1725–1731

Williams MS, Schreuder HT (2000) Guidelines for choosing volume equations in the presence of measurement error in height. Can J For Res 30:306–310

Williams MS, Wiant HV (1998) Double sampling with importance sampling to eliminate bias in tree volume estimation of the centroid method. For Ecol Manage 104:77–99

Williams RJ, Zerihun A, Montagu KD, Hoffmann M, Hutley LB, Chen X (2005) Allometry for estimating aboveground tree biomass in tropical and subtropical eucalypt woodlands: towards general predictive equations. Aust J Bot 53:607–619

Williams TM, Gresham CA (2006) Biomass accumulation in rapidly growing loblolly pine and sweetgum. Biomass Bioenergy 30:370–377

Williamson GB, Wiemann MC (2010) Measuring wood specific gravity . . . correctly. Am J Bot 97:519–524

Wilson TB, Meyers TP (2007) Determining vegetation indices from solar and photosynthetically active radiation fluxes. Agric For Meteorol 144:160–179

Witt GB, Harrington RA, Page MJ (2009) Is 'vegetation thickening' occurring in Queensland's mulga lands—a 50-year aerial photographic analysis. Aust J Bot 57:572–582

Wood GB, Wiant HV (1990) Estimating the volume of Australian hardwoods using centroid sampling. Aust For 53:271–274

Wood GB, Wiant HV (1992a) Test of application of centroid and importance sampling in a point-3P forest inventory. For Ecol Manage 53:107–115

Wood GB, Wiant HV (1992b) Test of centroid and importance sampling in an inventory of mature native eucalypt forest. In: Wood G, Turner B (eds) Integrating forest information over space and time. Proc. IUFRO conf., Jan 1992, Canberra. ANUTECH Pty Ltd, Canberra, pp 389–396

Wood GB, Wiant HV, Loy RJ, Miles JA (1990) Centroid sampling: a variant of importance sampling for estimating the volume of sample trees of radiata pine. For Ecol Manage 36:233–243

Wood MJ, Scott R, Volker PW, Mannes DJ (2008) Windthrow in Tasmania, Australia: monitoring prediction and management. Forestry 81:415–427

Woodall CW, Miles PD, Vissage JS (2005) Determining maximum stand density index in mixed species stands for strategic-scale stocking assessment. For Ecol Manage 216:367–377

Wunder J, Reineking B, Hillgarter FW, Bigler C, Bugmann H (2011) Long-term effects of increment coring on Norway spruce mortality. Can J For Res 41:2326–2336

Wutzler T, Wirth C, Schumacher J (2008) Generic biomass functions for Common beech (*Fagus sylvatica* L.) in Central Europe—predictions and components of uncertainty. Can J For Res 38:1661–1675

Wu Y, Guo L, Cui XH, Chen J, Cao X, Lin H (2014) Ground-penetrating radar-based automatic reconstruction of three-dimensional coarse root system architecture. Plant Soil 383:155–172

Wythers KR, Reich PB, Turner DP (2003) Predicting leaf area index from scaling principles: corroboration and consequences. Tree Physiol 23:1171–1179

Xiao C-W, Ceulemans R (2004a) Allometric relationships for below- and aboveground biomass of young Scots pine. For Ecol Manage 203:177–186

Xiao C-W, Ceulemans R (2004b) Allometric relationships for needle area of different needle age classes in young Scots pine: needles, branches and trees. Forestry 77:369–382

Ximenes FA, Gardner WD, Kathuria A (2008) Proportion of above-ground biomass in commercial logs and residues following the harvest of five commercial forest species in Australia. For Ecol Manage 256:335–346

Ximenes FA, Gardner WD, Richards GP (2006) Total above-ground biomass and biomass in commercial logs following the harvest of spotted gum (*Corymbia maculata*) forests of SE NSW. Aust For 69:213–222

Xing Z, Bourque CP-A, Swift DE, Clowater CW, Krasowski M, Meng F-R (2005) Carbon and biomass partitioning in balsam fir (*Abies balsamea*). Tree Physiol 25:1207–1217

Xu C-Y, Turnbull MH, Tissue DT, Lewis JD, Carson R, Schuster WSF, Whitehead D, Walcroft AS, Li J, Griffin KL (2012) Age-related decline of stand biomass accumulation is primarily due to mortality and not to reduction in NPP associated with individual tree physiology, tree growth or stand structure in a *Quercus*-dominated forest. J Ecol 100:428–440

Xue L, Pan L, Zhang R, Xu PB (2011) Density effects on the growth of self-thinning *Eucalyptus urophylla* stands. Trees 25:1021–1031

Yamamoto K (1994) A simple volume estimation system and its application to three coniferous species. Can J For Res 24:1289–1294

Yanai R, Battles J, Richardson A, Blodgett C, Wood D, Rastetter E (2010) Estimating uncertainty in ecosystem budget calculations. Ecosystems 13:239–48

Zakrzewski WT (2011) Estimating wood volume of the stem and branches of sugar maple (*Acer saccharum* Marsh.) using a stem profile model with implicit height. For Sci 57:117–133

Zakrzewski WT, Duchesne I (2012) Stem biomass model for jack pine (*Pinus banksiana* Lamb.) in Ontario. For Ecol Manage 279:112–120

Zakrzewski WT, MacFarlane DW (2006) Regional stem profile model for cross-border comparisons of harvested red pine (*Pinus resinosa* Ait.) in Ontario and Michigan. For Sci 52:468–475

Zapata-Cuartas M, Sierra CA, Alleman L (2012) Probability distribution of allometric coefficients and Bayesian estimation of aboveground tree biomass. For Ecol Manage 277:173–179

Zell J, Bösch B, Kändler G (2014) Estimating above-ground biomass of trees: comparing Bayesian calibration with regression technique. Eur J For Res 133:649–660

Zerihun A, Ammer C, Montagu KD (2007) Evaluation of a semi-empirical model for predicting fine root biomass in compositionally complex woodland vegetation. Ann For Sci 64:247–254

Zerihun A, Montagu KD (2004) Belowground to aboveground biomass ratio and vertical root distribution responses of mature *Pinus radiata* stands to phosphorus fertilization at planting. Can J For Res 34:1883–1894

Zerihun A, Montagu KD, Hoffman MB, Bray SG (2006) Patterns of below- and aboveground biomass in *Eucalyptus populnea* woodland communities of northeast Australia along a rainfall gradient. Ecosystems 9:501–515

Zhang Y, Borders BE, Bailey RL (2002) Derivation, fitting, and implication of a compatible stem taper-volume-weight system for intensively managed, fast growing loblolly pine. For Sci 48:595–607

Zhang Y, Chen J, Miller JR (2005) Determining digital hemispherical photograph exposure for leaf area index estimation. Agric For Meteorol 133:166–181

Zhu SP, Huang CL, Su Y, Sato M (2014) 3D ground penetrating radar to detect tree roots and estimate root biomass in the field. Remote Sens 6:5754–5773

Zianis D (2008) Predicting mean aboveground forest biomass and its associated variance. For Ecol Manage 256:1400–1407

Zianis D, Mencuccini M (2004) On simplifying allometric analyses of forest biomass. For Ecol Manage 187:311–332

Zubritsky E (2014) New NASA probe will study earth's forests in 3D. http://www.pddnet.com/news/2014/09/new-nasa-probe-will-study-earths-forests-3d. Accessed 18 Feb 2015

Index

A

Abies
  A. alba, 161
  A. cilicica, 44–46
Accuracy of the survey, 141
Acer saccharum, 43, 146
Aerial, 109, 132
Aerial photography, 117, 154–159
Africa, 3, 17, 47, 58, 63, 64, 150
African juniper. *See Juniperus procera*
Airborne, 111, 117, 121, 145, 154, 158–161
Airborne remote sensing. *See* Remote sensing
Aircraft, 1, 106, 123, 143, 145–146, 154, 155, 157–159
Aleppo pine. See Pinus, P. halepensis
Allometry, 61
Amazon Basin, 64, 133
America. *See* United States of America
American, 2, 35, 62, 89, 112, 123, 162
Angle count sampling. *See* Point sampling
Angle gauge sampling. *See* Point sampling
Asia, 3, 63, 64
Australia, 2, 13, 14, 18, 22, 30, 31, 57, 62, 65, 69, 85, 87, 90, 91, 93, 100, 117, 150, 151, 158–160, 163
Austria, 131

B

Bacteria, 59
Bark
  gauge, 18
  thickness, 17–18, 32, 33, 147

Basal area
  factor, 77, 78, 80, 88, 125, 128, 129
  prism, 77, 78
  stand, 71, 72, 74, 76–79, 81, 83, 88, 93, 95, 128, 129, 162
  wedge, 77
Basic density. *See* Wood, density
Bavaria, Germany, 161
Beer–Lambert law, 149
Belgium, 92
Bias, 5–10, 13, 29, 31, 35, 39, 64, 80, 98, 125, 126, 130, 147
Bioenergy, 12, 54, 94
Biomass
  estimation functions, 57, 60–70, 88–91, 131, 156
  expansion factors, 67, 68, 89, 90
  functions, 62, 64, 65, 68–70, 89, 131
  stand, 71, 72, 74, 81, 88–91, 117, 131, 163
Bitterlich sampling. *See* Point sampling
Blackbutt. *See Eucalyptus, E. pilularis*
Bootstrap, 114, 117
Boreal regions, 70, 91
Borneo, 64
Branching, 3, 8, 12, 13, 16, 19, 26, 31, 33, 45, 47, 48, 53–56, 59–61, 65, 67, 126, 147, 149, 150, 158
Britain, 162
Brutian pine. *See Pinus, P. brutia*
Bumps, 16, 31
Bursera simaruba, 39
Buttress, 12, 13, 16, 17
Butt swell, 29–31, 34, 44, 45